EAI/Springe Innovations in Communication and Comp ing

Series Ed tac, European Alliance for Innovation, Ghent, Belgium
Imrich

The impact of information technologies is creating a new world yet not fully understood. The extent and speed of economic, life style and social changes already perceived in everyday life is hard to estimate without understanding the technological driving forces behind it. This series presents contributed volumes featuring the latest research and development in the various information engineering technologies that play a key role in this process. The range of topics, focusing primarily on communications and computing engineering include, but are not limited to, wireless networks; mobile communication; design and gaming; interaction; e-health and pervasive healthcare; energy management; grids; internet of things; cognitive radio networks; computation; cloud computing; ubiquitous connectivity, and in mode general smart living, smart cities, Internet of Things and more. The series publishes a combination of expanded papers selected from hosted and sponsored European Alliance for Innovation (EAI) conferences that present cutting edge, global research as well as provide new perspectives on traditional related engineering fields. This content, complemented with open calls for contribution of book titles and individual chapters, together maintain Springer's and EAI's high standards of academic excellence. The audience for the books consists of researchers, industry professionals, advanced level students as well as practitioners in related fields of activity include information and communication specialists, security experts, economists, urban planners, doctors, and in general representatives in all those walks of life affected ad contributing to the information revolution.

Indexing: This series is indexed in Scopus, Ei Compendex, and zbMATH.

About EAI - EAI is a grassroots member organization initiated through cooperation between businesses, public, private and government organizations to address the global challenges of Europe's future competitiveness and link the European Research community with its counterparts around the globe. EAI reaches out to hundreds of thousands of individual subscribers on all continents and collaborates with an institutional member base including Fortune 500 companies, government organizations, and educational institutions, provide a free research and innovation platform. Through its open free membership model EAI promotes a new research and innovation culture based on collaboration, connectivity and recognition of excellence by community.

Dalai Tang • Joni Zhong • Dalin Zhou
Editors

Mobile Wireless Middleware, Operating Systems and Applications

10th International Conference on Mobile Wireless Middleware, Operating Systems and Applications (MOBILWARE 2021)

Springer

EAI
RESEARCH MEETS INNOVATION

Editors
Dalai Tang
Inner Mongolia University of Finance and Economics
Hohhot, China

Joni Zhong
Hong Kong Polytechnic University
Hung Hom, Hong Kong

Dalin Zhou
University of Portsmouth
Portsmouth, UK

ISSN 2522-8595 ISSN 2522-8609 (electronic)
EAI/Springer Innovations in Communication and Computing
ISBN 978-3-030-98670-4 ISBN 978-3-030-98671-1 (eBook)
https://doi.org/10.1007/978-3-030-98671-1

This Springer imprint is published by the registered company Springer Nature Switzerland AG
The registered company address is: Gewerbestrasse 11, 6330 Cham, Switzerland

Preface

We are delighted to introduce the proceedings of the 2021 European Alliance for Innovation (EAI) International Conference on Mobile Wireless Middleware, Operating Systems, and Applications (MobileWare 2021). This conference has brought researchers, developers, and practitioners around the world to contribute to organized topics of 5G wireless communication, wireless sensor networks, knowledge extraction, instantaneous availability, complex networks, computer vision for mobile application, and mobile support robots. The state-of-the-art research conducted both theoretically and experimentally was presented.

The proceedings consisted of 12 papers after rigorous peer review. All papers were orally presented during the conference virtually in a live stream. The conference program arranged three keynote speeches. We would like to express our sincere appreciation and acknowledgment to the distinguished keynote speakers: Assoc. Prof. Janos Botzheim from Eötvös Loránd University, Hungary; Assoc. Prof. Celimuge Wu from the University of Electro-Communications, Japan; and Prof. Naoyuki Kubota from Tokyo Metropolitan University, Japan.

Here, we would also like to express our sincere thanks to all individuals who supported MobilWare 2021. All program committee chairs and members are appreciated for their rigorous review of the submitted papers and their successful organizational work. We convey our highest appreciation and gratitude to all the authors and participants of this conference who made it happen during the tough time of the Covid-19 pandemic. Special thanks are extended to Prof. Imrich Chlamtac for steering the conference, and to Lenka Lezanska and Eliska Vlckova from EAI for their kind support throughout the publication of the conference proceedings.

We hope that researchers, engineers, and students who share similar interests in the related topics of MobilWare 2021 will find the conference proceedings useful.

Hohhot, China — Dalai Tang
Hung Hom, Hong Kong — Joni Zhong
Portsmouth, UK — Dalin Zhou

Conference Organization

Steering Committee
Imrich Chlamtac, University of Trento, Italy

Organizing Committee

General Chairs
Fanyu Bu, Inner Mongolia University of Finance and Economics, China
Bo Gao, Inner Mongolia University of Finance and Economics, China
GongFa Li, Wuhan University of Science and Technology, China

Technical Program Committee Chairs
Liang Zhao. Dalian University of Technology, China
Dalai Tang, Inner Mongolia University of Finance and Economics, China

Web Chairs
Jinseok Woo, Tokyo University of Technology, Japan
Xuezheng Yue, University of Shanghai for Science and Technology, China

Publicity and Social Media Chairs
Yinfeng Fang, Hangzhou Dianzi University, China
Yuichiro Toda, Okayama University, Japan

Workshops Chairs
Haishan Bao, Inner Mongolia University of Finance and Economics, China
Wuyungerile Li, Inner Mongolia University of Finance and Economics, China
Zhong-Zhou Lan, Inner Mongolia University of Finance and Economics, China
Yana A, Inner Mongolia University of finance and economics, China

Sponsorship & Exhibits Chair
Suyalatu Dong, Inner Mongolia University of Finance and Economics, China
Song Yang, Beijing Institute of Technology, China

Publication Chairs

Dalin Zhou, University of Portsmouth, United Kingdom
Junpei Zhong, The Hong Kong Polytechnic University

Panels Chairs

Baojun Sun, Inner Mongolia University of Finance and Economics, China
Tong Cui, Shenyang Aerospace University, China

Tutorials Chairs

Ting Wang, Shandong University of Science and Technology, China
Linlin Xu, Inner Mongolia University of Finance and Economics, China

Demos Chairs

Peng Zhang, Ningxia University, China
Xinwei Zhang, Inner Mongolia University of finance and economics, China

Posters and PhD Track Chairs

Yubo Guo, Inner Mongolia University of Finance and Economics, China
Hua Wu, Inner Mongolia University of finance and economics, China

Local Organisation Chairs

Pingquan Wang, Inner Mongolia University of Finance and Economics, China
Ming Jing Du, Inner Mongolia University of finance and economics, China

Technical Program Committee

Liang Zhao, Dalian University of Technology, China
Dalai Tang, Inner Mongolia University of finance and economics, China
Fanyu Bu, Inner Mongolia University of finance and economics, China
Wuyungerile Li, Inner Mongolia University, China
Pingquan Wang, Inner Mongolia University of finance and economics, China
GongFa Li, Wuhan University of Science and Technology, China
Junpei Zhong, The Hong Kong Polytechnic University, China
Xuezheng Yue, University of Shanghai for Science and Technology, China
Yinfeng Fang, Hangzhou Dianzi University, China
Yuichiro Toda, Okayama University, Japan
Haishan Bao, Inner Mongolia University of finance and economics, China
Dalin Zhou, University of Portsmouth, UK
Tong Cui, Shenyang Aerospace University, China
Ting Wang, Shandong University of Science and Technology, China
Jinseok Woo, Tokyo University of Technology, Japan
Bo Gao, Inner Mongolia University of finance and economics, China
Zhong-Zhou Lan, Inner Mongolia University of finance and economics, China
Baojun Sun, Inner Mongolia University of finance and economics, China
Linlin xu, Inner Mongolia University of finance and economics, China
Yubo Guo, Inner Mongolia Agricultural University, China
Peng Zhang, Ningxia University, China
Suyalatu Dong, Inner Mongolia University of finance and economics, China

Contents

Human Behavior Estimation Using Micro-vibration Sensor Based on Deep Boltzmann Machine

Naoki Doteguchi, Shuai Shao, and Naoyuki Kubota

1 Introduction

The aging of the population in developed countries, including Japan, has become a major social issue. According to a survey by the World Health Organization, it is predicted that by 2050, the population over 65 years old will account for 28% of the total population [1]. In this situation, domestic accidents among the elderly are becoming a significant problem. These accidents have a great impact on the health of the elderly. To prevent and detect such accidents, a monitoring system is needed [2, 3]. To achieve this, it is necessary to understand the behavior of the person being monitored accurately. One of the most popular methods for this purpose is to use camera images for behavior estimation. However, one of the significant problems in using camera images is the issue of privacy. In addition, the measurement of daily life using cameras is considered stressful for the elderly [4–6].

Table 1 shows the comparison of some sensors that can be used in smart homes. Considering the basic requirements for privacy protection and non-wearing, we mainly choose from various indirect motion detect sensors. The lateral comparison found that the vibration sensor can meet all needs. But at the same time, we must admit that indirect data sacrifice the advantage of accuracy, which is the main problem that the research must solve.

Therefore, this paper aims to develop a behavior estimation system using micro-vibration sensors. However, one of the problems with micro-vibration sensors is that the measured data contains a lot of noise and has different feature values depending on the behavioral state and environmental conditions. Therefore, this paper proposes

N. Doteguchi (✉) · S. Shao · N. Kubota
Tokyo Metropolitan University, Tokyo, Japan
e-mail: doteguchi-naoki@ed.tmu.ac.jp; shao-shuai1@ed.tmu.ac.jp; kubota@ed.tmu.ac.jp

D. Tang et al. (eds.), *Mobile Wireless Middleware, Operating Systems and Applications*, EAI/Springer Innovations in Communication and Computing,
https://doi.org/10.1007/978-3-030-98671-1_1

Table 1 Advantages and disadvantages of different sensors

	Wearable sensor	Vision sensor	Indirect motion detect sensors			
	Body-sensor	Camera	Doppler radar	Infrared sensor	Vibration mat	Vibration sensor
Ease-to-set	–	+++	+++	+	+	+++
Privacy protection	–	–	++	++	++	+++
Cheap	–	–	++	+	–	+++
Low-energy	+	–	++	++	+++	++
Imperceptible	–	+	++	+++	++	+++
Stable	+++	+++	++	++	+++	++
Accuracy	+++	+++	++	++	++	+
Multifunction	+++	+++	+	+	+	++

a system that removes noise and extracts features by using a deep Boltzmann machine to estimate behavior from micro-vibration sensor data and confirm its effectiveness.

2 Related Method

2.1 *Boltzmann Machine*

Boltzmann machine [7] is a neural network model that was proposed in the 1980s. It is classified as a generative model because the behavior of the network is described probabilistically. The Boltzmann machine is a model with a graph structure, as shown in Fig. 1. Each node is divided into viewing nodes v_i and hidden nodes h_j, each of which takes the value 0 or 1. Since the Boltzmann machine is an undirected graph, each joint weight is $w_{ji}(i \neq j)$, $w_{ii} = 0$. If the bias at each node is b_i, then the energy of the Boltzmann machine is expressed by the following equation. The following equation expresses the energy of the Boltzmann machine. Here X is a vector of v and h.

$$\phi\left(X;\theta\right) = -\sum_{i} b_i x_i - \sum_{ij} w_{ij} x_i x_j \tag{1}$$

From Eq. (1), the generation probability of the Boltzmann machine is defined as follows:

$$p\left(V = v|\theta\right) = \frac{1}{Z\left(\theta\right)} \exp\left\{-\phi\left(v;\theta\right)\right\} \tag{2}$$

$$Z\left(\theta\right) = \sum_{v} \exp\left\{-\phi\left(v;\theta\right)\right\} \tag{3}$$

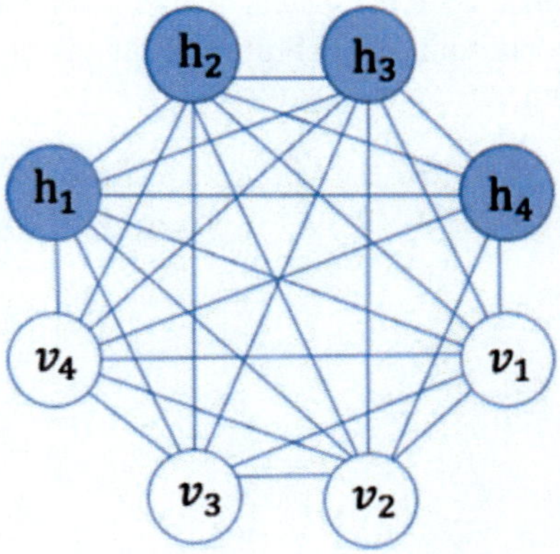

Fig. 1 Boltzmann machine with hidden variables

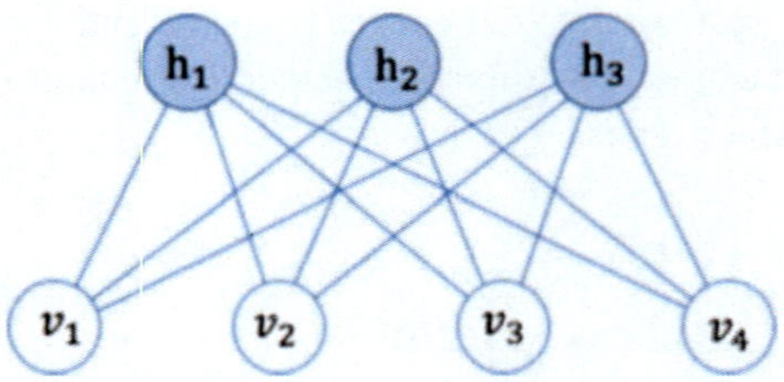

Fig. 2 Restricted Boltzmann machine

2.2 Restricted Boltzmann Machine

The Boltzmann machine takes a very long time to train due to the combinatorial explosion, a significant problem in practice. We use the restricted Boltzmann machine [8] to solve this problem, a model that eliminates the coupling between the visible and hidden layers (Fig. 2). The energy function of the restricted Boltzmann machine is as follows:

$$\phi(v, h; \theta) = -\sum_{ij} v_i w_{ij} h_j - \sum_i b_i v_i - \sum_j c_j h_j \tag{4}$$

The probability model can be expressed as follows:

$$p(V = v, H = h|\theta) = \frac{1}{Z(\theta)} \exp\{-\phi(v, h; \theta)\} \tag{5}$$

$$Z(\theta) = \sum_v \exp\{-\phi(v, h; \theta)\} \tag{6}$$

The restricted Boltzmann machine is a bipartite graph, and one of its essential properties is the conditional independence of each variable. In other words, once the probability distribution of one layer is determined, the probability distribution of the other layer can be calculated independently. The following equation can obtain the conditional probabilities of the visible and hidden layers. Here sig () is the sigmoid function:

$$p(v_i|h; \theta) = \mathrm{sig}\left(\sum_j w_{ij} h_j - b_j\right) \tag{7}$$

$$p\left(h_j|v; \theta\right) = \mathrm{sig}\left(\sum_i w_{ij} v_i - c_i\right) \tag{8}$$

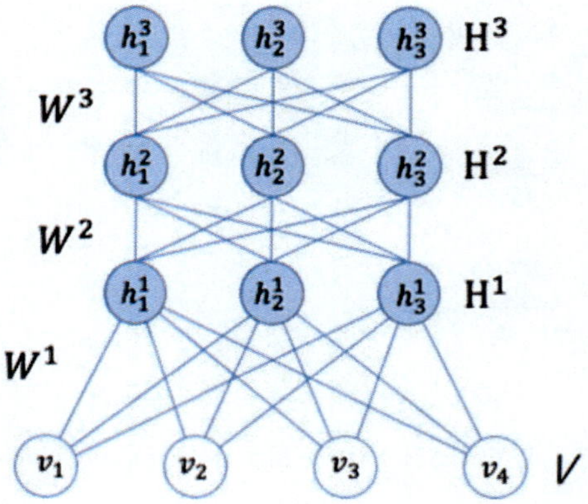

Fig. 3 Deep Boltzmann machine

This makes it easy for the restricted Boltzmann machine to sample alternately according to the conditional probabilities of each stratum.

2.3 *Deep Boltzmann Machine*

Deep Boltzmann machine [9] extends the restricted Boltzmann machine with a deep graph structure to learn more complex probability distributions. The structure of a deep Boltzmann machine is an extended model of restricted Boltzmann machines. The bottom layer is the visible layer, and the other layers are hidden layers (Fig. 3). The energy of the deep Boltzmann machine is defined by the following Eq. (9). In addition, there is a problem that deep Boltzmann machines tend to fall into local minimum depending on the initial values of each parameter. In order to avoid this problem, it is necessary to select appropriate initial values for each parameter. To avoid this problem, we pre-train the deep Boltzmann machine as a restricted Boltzmann machine by extracting two layers at a time and then train the entire model:

$$\phi\,(v, h; \theta) = -\sum_{i\in V}\sum_{j\in H^{(1)}} w_{ij}^{(1)} v_i h_j^{(1)} - \sum_{r=2}^{R}\sum_{j\in H^{(r-1)}}\sum_{i\in H^{(r)}} w_{ij}^{(r)} h_i^{(r-1)} h_j^{(r)} \tag{9}$$

3 Human Behavior Estimation System

In the proposed system, we chose to place two sensors in the same room as sensors to measure the daily activities of elderly people living alone. The vibration sensor does not use images or 3D distance, which can protect the privacy of the elderly

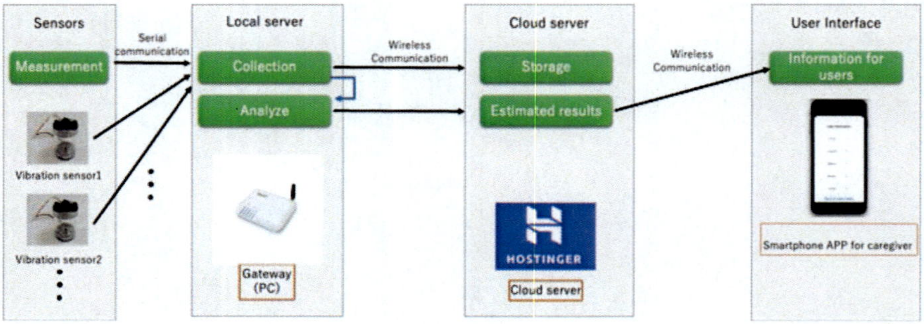

Fig. 4 Workflow of the human behavior estimation system

to the greatest extent. Therefore, the proposed system can reduce the psychological burden of the elderly.

Figure 4 shows the workflow of this system [10]. The real-time data collected by each sensor is sent to the processing terminal through serial communication. The data processing terminal merges and processes the data of the two sensors according to the algorithm in the following text and outputs the user's location and current state. The results are uploaded to the cloud server in real time via Wi-Fi. Caregivers can view the current status of the elderly through their smartphones synchronized with the cloud server in real time.

4 Behavior Estimation Algorithm

This system estimates the behavioral state by learning a deep Boltzmann machine and a neural network using the data measured by two micro-vibration sensors. The flowchart of the system is shown in Fig. 5.

First, we use information from micro-vibration sensors installed at two different locations. The data obtained from these sensors are processed by LPF, WRMS, and ACF to convert them to 0 and 1. After that, we estimate the target's behavior by extracting features using a deep Boltzmann machine and classifying them using a neural network.

4.1 Low-Pass Filter (LPF)

The measured value by the micro-vibration sensor contains a lot of noise. Therefore, the noise is removed by using an LPF that cuts the signal above a specific frequency. Suppose the output is y_i and the input is x_i, the following equation is obtained:

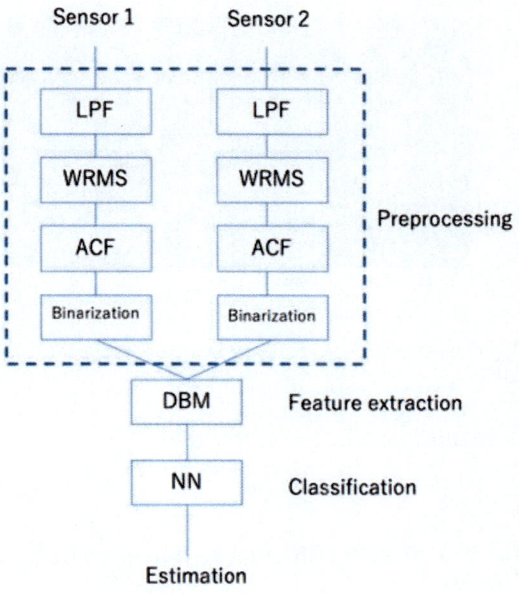

Fig. 5 Human behavior estimation system

$$y_i = ax_i + (1 - a)\, y_{i-1} y_i \tag{10}$$

In this case, the cutoff frequency is $\omega_c = \alpha/(1 + \alpha) \cdot T_s$, where T_s is the sampling frequency of the data. In this system, $\alpha = 0.5$ is used.

4.2 *Weighted Root Mean Square (WRMS)*

However, if the degree of smoothing is increased, the amplitude becomes smaller, and the feature values are lost. Therefore, weighted RMS is used to extract feature values while maintaining the amplitude. Since the waveform data is affected by the time series at a certain time, WRSM (σ_t) is calculated using the data y_i at the time $(t, t-1, t-2, \ldots, t-n)$:

$$\sigma_t = \sqrt{\frac{\sum_{i=0}^{n} w_i \delta_i^2}{n \sum_{i=0}^{n} w_i}} \tag{11}$$

$$\delta_i = y_i - \overline{y} \tag{12}$$

$$w_i = \mathrm{e}^{-(n/2-i)^2 \cdot \beta} \tag{13}$$

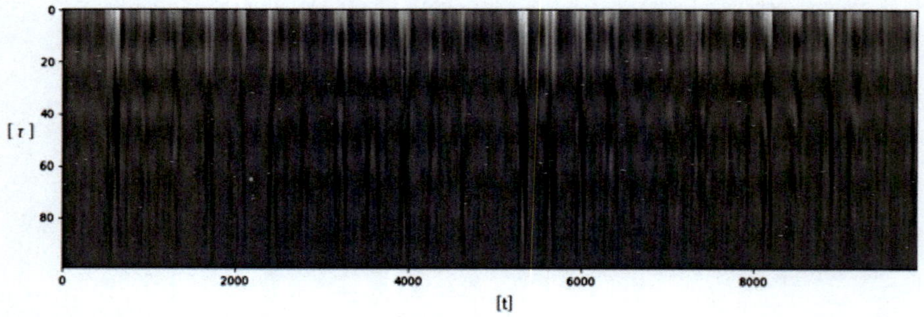

Fig. 6 Spectrogram with the autocorrelation function

In this system, $t = 24$, $\beta = 0.01$.

4.3 Autocorrelation Function (ACF)

The autocorrelation function measures how well a signal at a given time matches a movement that has shifted itself in time and is expressed as a function of the magnitude of the time shift. Human behavior, such as walking, has a cyclic nature. Therefore, it is considered that the autocorrelation function can be used to estimate the behavior by extracting the features with periodicity.

The autocorrelation function AC_τ of the time interval τ at a certain time t can be expressed as follows:

$$AC_\tau = \frac{1}{N} \sum_{t=0}^{N-1} \sigma_t \overline{\sigma}_{t-\tau} \tag{14}$$

In this system, the measurement data is 50 Hz and $N = 100$ to secure the information volume of 2 s in the time direction. The spectrogram obtained by the autocorrelation function is shown in Fig. 6.

4.4 Binarization

Each node of the Boltzmann machine can take only 0 or 1 values. Therefore, the spectrogram obtained by the autocorrelation function cannot be directly used as the input of the Boltzmann machine. Consequently, it is necessary to convert it into a form suitable for input.

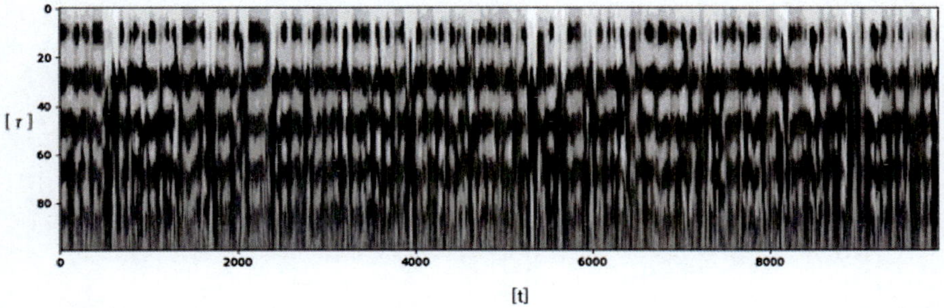

Fig. 7 Input to Boltzmann machine

In this system, ten nodes are prepared as inputs for time t and interval τ, and they are transformed into ten levels of inputs based on the magnitude of the signal. The system uses logarithms to convert the information in the small amplitude part more closely. The output is y, and the input is x, as shown in the following equation:

$$y = \mathrm{Round}\left(1.8\log\left(10^2\left(1 - 10^{2.5}\right)|x| - 1\right)\cdot\frac{x}{|x|}\right) \tag{15}$$

Using the above transformation, the input to the Boltzmann machine is as follows (Fig. 7).

4.5 *Feature Extraction by DBM*

In our system, the Boltzmann machine is expected to play two major roles: the first role is to extract the features of the input as an autoencoder. The second role is to absorb and remove noise from the input using the Boltzmann machine's ability to remember the learned input and recall it when given a similar input (Fig. 8).

4.6 *Classification by Neural Network*

The features extracted by the Boltzmann machine are used to estimate the behavior using an all-connecting neural network. A logistics function is used as the activation function (Fig. 9).

In this system's state estimation by the neural network, the state corresponding to the node that shows the largest output among the output layer is assumed to be the output. To avoid false judgments, a threshold value is set. Suppose the difference between the most significant output and the second largest output is less than the

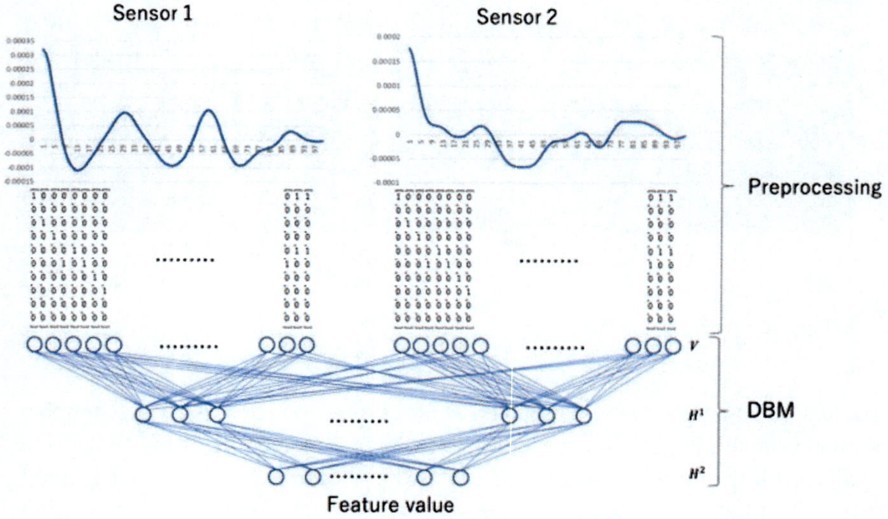

Fig. 8 Feature extraction by DBM

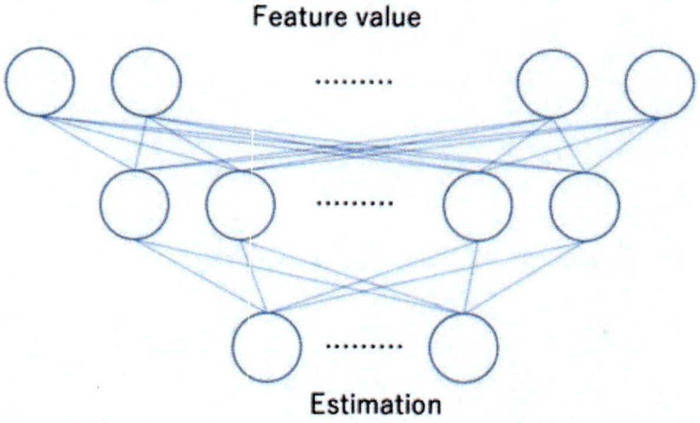

Fig. 9 Classification by neural network

threshold value. In that case, the system outputs a message stating that the degree of separation is low and that it cannot discriminate.

4.7 *Time Series Majority Decision*

However, these data are time series data, and the behavior patterns retain such time series information. Therefore, by performing majority voting in the direction of the time series, it is possible to estimate behavior that considers the time series data.

5 Human Behavior Estimation Experiment

5.1 Training the Proposed Model

Two micro-vibration sensors were placed on the floor at an interval of 2.4 m, and the actor behave between the sensors was measured. The measured behavioral patterns are summarized in Table 2. For each behavioral pattern, we used 3000 pieces of data, 15,000 pieces in total, as training data. The parameters of the deep Boltzmann machine and neural network are shown in Tables 3 and 4. We prepared five thresholds of 0, 0.05, 0.1, 0.2, and 0.3 and examined the estimation results for each of them. Figure 10 shows the confusion matrix of the estimation results by each threshold, and Table 5 shows the normalized estimation accuracy excluding the indistinguishable ones.

The results of the behavior estimation are shown in Fig. 10, and the accuracy is 87.6% at the threshold of 0. The accuracy is 87.6% at the threshold value of 0, which means that we can estimate the behavior with high accuracy. In addition, some misjudgments, such as walk1 and walk2, can be attributed to the fact that they have similar characteristics to the original behavior. Table 5 shows that the estimation accuracy without the discriminator is 87.6%, while the accuracy increases as the threshold is increased. On the other hand, as the threshold value was increased, the amount of data that could not be discriminated against increased. At the threshold value of 0.3, about 10% of the data could not be discriminated against.

Table 2 Behavioral pattern

No.	Behavioral pattern	Remarks
0	walk1	Normal walk
1	walk2	Walk with a limp
2	jump	Jumping
3	jog	Jogging
4	no_signal	No behavior

Table 3 Parameters of the deep Boltzmann machine

	Number of nodes
First layer	2000
Second layer	1000
Third layer	500

Table 4 Parameters of the neural network

	Number of nodes
First layer	500
Second layer	100
Third layer	5

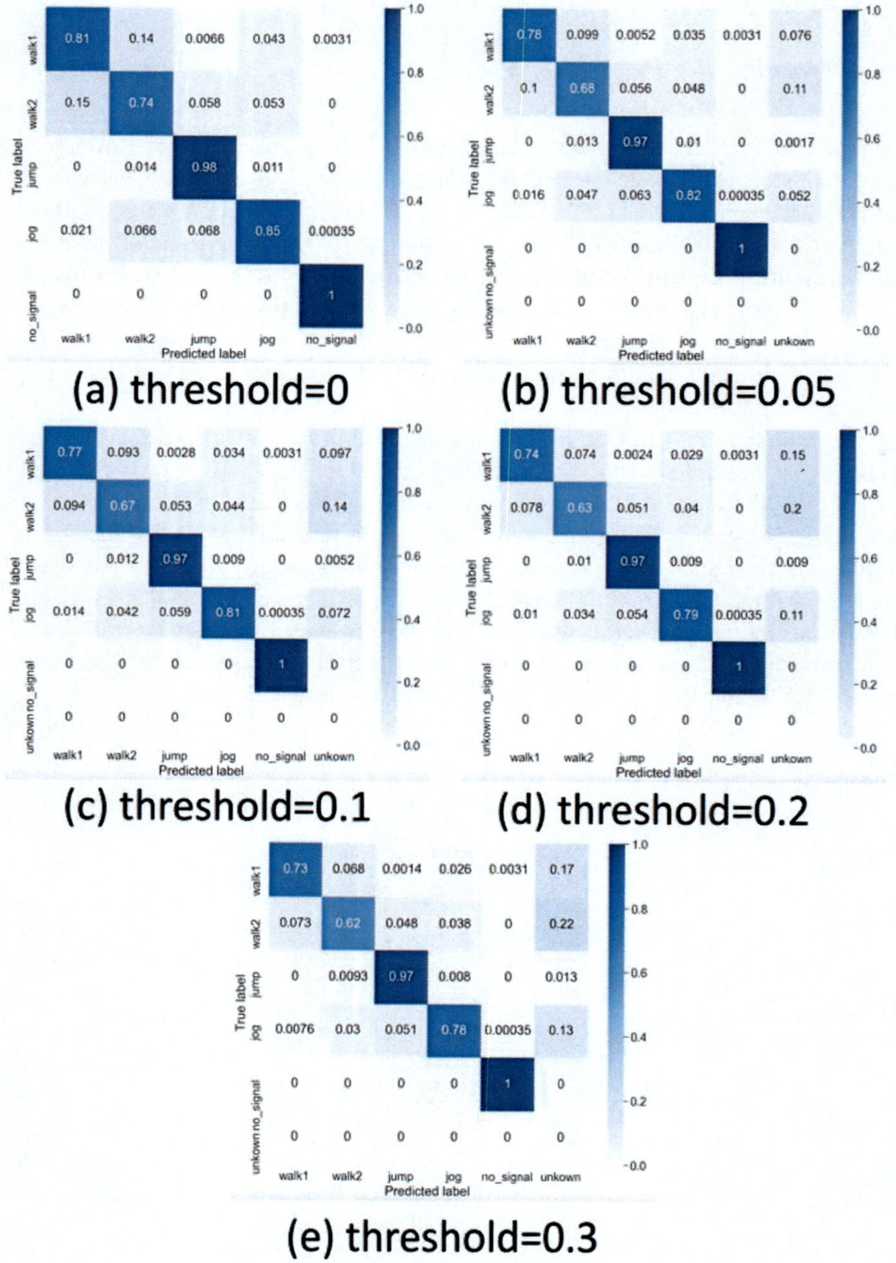

Fig. 10 Result of training

Table 5 Normalized estimation accuracy

	Behavior estimation					
Threshold	0:walk1	1:walk2	2:jump	3:jog	4:no_signal	Average
0	0.810	0.740	0.980	0.850	1.000	0.876
0.05	0.844	0.764	0.972	0.865	1.000	0.889
0.1	0.853	0.779	0.975	0.873	1.000	0.896
0.2	0.871	0.788	0.979	0.888	1.000	0.905
0.3	0.880	0.795	0.983	0.897	1.000	0.911

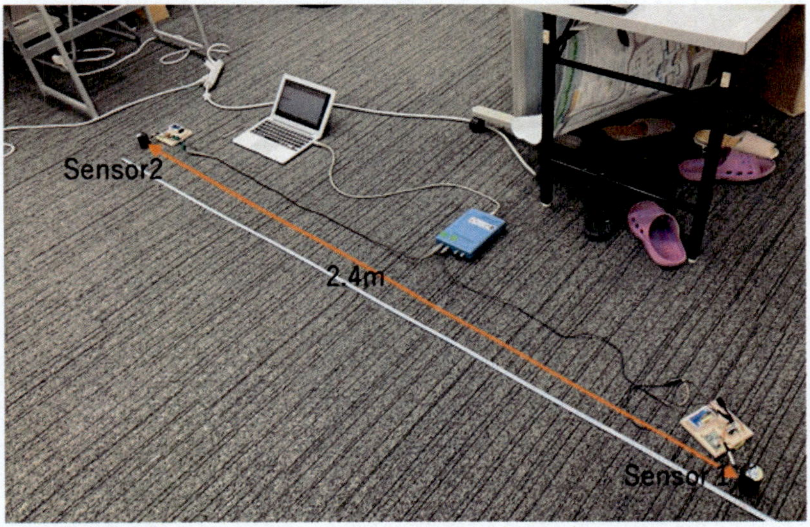

Fig. 11 Scene of the experiment

Fig. 12 Behavioral state transition of the measurement target

5.2 *Experimental Methods*

In experimenting, the behavioral state transitions of the measurement target were defined, as shown in Fig. 12. In this experiment, the behavioral state transitions were performed continuously, with no behavior in between each of the four behavioral states the trained behavior estimation system in 4.1. The measurement method is the same as in 4.1, where the sensor is placed interval of 2.4 m, and the subject walks back and forth near the sensor (Figs. 11 and 12).

Table 6 Estimation accuracy

Threshold	Accuracy
0	0.696
0.05	0.701
0.1	0.702
0.2	0.718
0.3	0.720

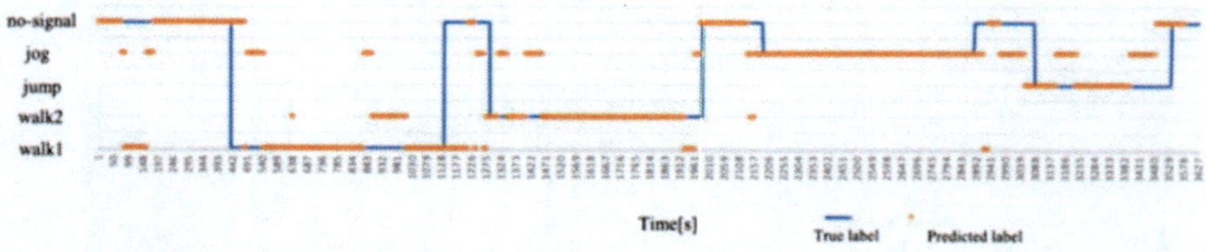

Fig. 13 Results of behavior estimation (threshold = 0)

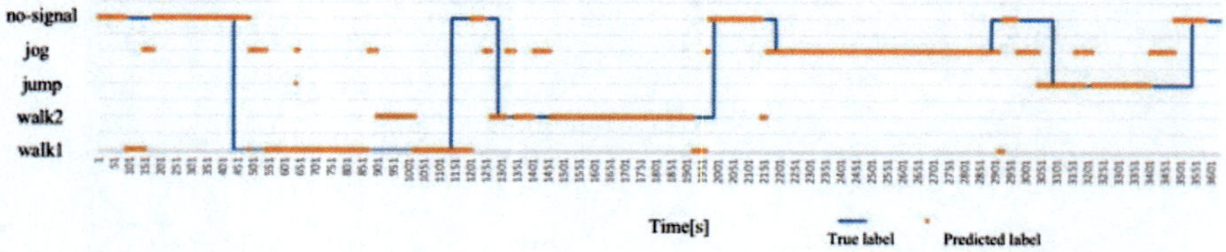

Fig. 14 Results of behavior estimation (threshold = 0.3)

5.3 Behavior Estimation Results

Table 6 shows the results of behavior estimation for the threshold values of 0, 0.05, 0.1, 0.2, and 0.3. Figure 13 shows the time series of the behavior estimation for the threshold values of 0 (no discrimination) and 0.3 (Fig. 14).

The estimation accuracy was 69.6% at threshold 0 and 72% at threshold 0.3. For each state, walk1 was estimated as walk2 in some cases, and walk2 was estimated as jog. Jog was estimated with considerable accuracy. For no_signal, we estimated it with relatively high accuracy in 4.1, but many parts were not estimated well in this experiment.

5.4 Consideration

As for the estimation result of no_signal, the training data used for this state was almost no input. The output of the autocorrelation function was minimal and almost flat at the time of transformation to 10 levels. Therefore, when some input information appeared regardless of noise, the output was not no_signal but other behavioral states. The reason walk1 is often misidentified as walk2 because walk1 and walk2 are both walking behaviors and the degree of separation between them is low, making it difficult to estimate. The reason jump and jog are misestimated is that the frequency of these two behavior patterns is close to each other.

6 Conclusion

In this paper, we proposed a behavior estimation system using micro-vibration sensors and a deep Boltzmann machine as a behavior estimation method that did not rely on camera images in consideration of privacy issues. We confirmed the possibility of the proposed method by training the system for five behavior patterns. In addition, we confirmed the effectiveness of the proposed method by estimating the behaviors when these patterns were performed continuously. As future work, we need to improve the state of no action in this experiment. In addition, we would like to improve the estimation accuracy of the proposed system by combining it with multiple sensors, although we used only a micro-vibration sensor.

References

1. Muramatsu, N., Akiyama, H.: Japan: super-aging society preparing for the future. Gerontologist. **51**(4), 425–432 (2011)
2. Panatto, D., Gasparini, R.: Survey of domestic accidents in the elderly in the province of Genoa (northern Italy). J. Prev. Med. Hyg. **50**(1), 53–57 (2009)
3. Oyetunji, T.A., Ong'uti, S.K.: Epidemiologic trend in elderly domestic injury. J. Surg. Res. **173**(2), 206–211 (2012)
4. Caine, K.E., Fisk, A.D., Rogers, W.A.: Benefits and privacy concerns of a home equipped with a visual sensing system: a perspective from older adults. Proceedings of the Human Factors and Ergonomics Society Annual Meeting, 2006
5. Demiris, G., et al.: Older adults' privacy considerations for vision based recognition methods of eldercare applications. Technol Health Care. **7**(1), 41–48 (Jan. 2009)
6. Fuse, H., Isshikim, M.: Proposal of life monitoring system for the elderly using voice interactive robot. IEEE 9th Global Conference on Consumer Electronics (GCCE), 2020
7. Ackley, D.H., Hinton, G.E., Sejnowski, T.J.: A learning algorithm for Boltzmann machines. Cogn. Sci. **9**, 147–169 (1985)
8. Smolensky, P.: Information processing in dynamical systems: foundations of harmony theory. In: Parallel Distributed Processing: Explorations in the Microstructure of Cognition, vol. 1, pp. 194–281. MIT Press, Cambridge (1986)
9. R. Salakhutdinov, G.E. Hinton: Deep Boltzmann machines. Proceedings of the Twelfth International Conference on Artificial Intelligence and Statistics, pp. 448–455 (2009)
10. Shao, S., Yamamoto, K., Kubota, N.: An elderly monitoring system based on multiple ultra-sensitive vibration and pneumatic sensors. J. Adv. Comput. Intell. Intell. Inform. **25**(4), 423–431 (2021)

Topological Tracking for Mobility Support Robots Based on Multi-scale Batch Learning Growing Neural Gas

Naoki Doteguchi and Naoyuki Kubota

1 Introduction

Recently, the basic concept of cyber-physical systems has been extended to various research fields such as cyber-physical-social systems (CPSS) and cyber-physical-human systems (CPHS). One of the original ideas on CPSS [1] follows three interacting worlds: the physical (World One), mental (World Two), and artificial (World Three), according to Karl Popper's theory. Thanks to the rapid progress of network technology, we can realize real-time measurement of people, moving objects, and environments in a wide area. However, it is difficult to simulate and analyze the interaction in these three worlds because we must deal with qualitatively different big data simultaneously [2]. So, it is often necessary to extract features and structures based on graph theory and topology to reproduce real-world phenomena in the cyber world [3, 4]. Therefore, we proposed the concept of a topological twin [5, 6].

With topological twins, we can achieve (1) and discover topological structures that are not easily detectable in the real world; (2) reproduce the topological structures in the display by mathematical methods under the virtual world; and (3) analyze and predict the real world through simulation in the virtual world. Figure 1 shows the concept of topological twin in cyber-physical-social systems. We can extract topological features from big data as topological big data at various levels.

Furthermore, we need a multi-scopic approach to deal with inference, learning, search, and prediction based on topological and graphical data as the methodology of topological intelligence. While we have to deal with the physical dynamics at the microscopic level, we have to deal with spatiotemporal qualitative relation-

N. Doteguchi (✉) · N. Kubota
Tokyo Metropolitan University, Tokyo, Japan
e-mail: doteguchi-naoki@ed.tmu.ac.jp; kubota@tmu.ac.jp

D. Tang et al. (eds.), *Mobile Wireless Middleware, Operating Systems and Applications*, EAI/Springer Innovations in Communication and Computing,
https://doi.org/10.1007/978-3-030-98671-1_2

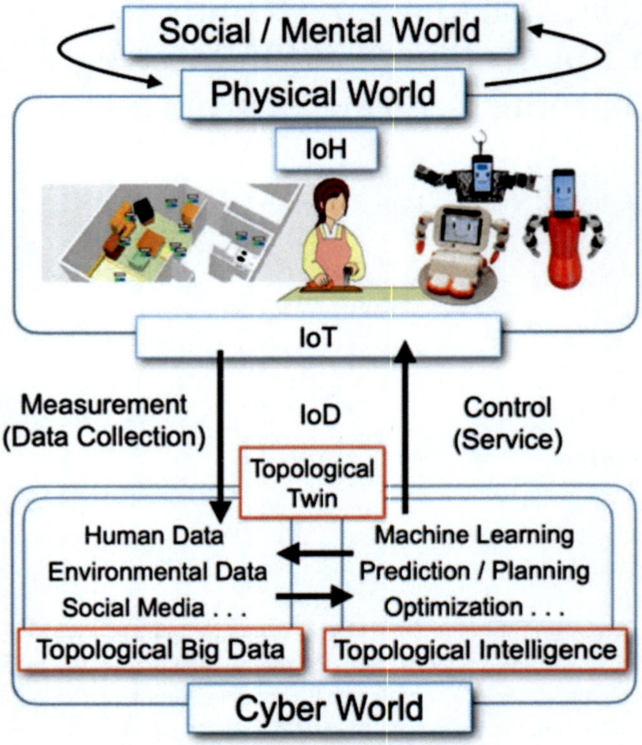

Fig. 1 Topological twin in cyber-physical-social systems

ships between objects, people, culture, and knowledge at the macroscopic level. Furthermore, we need a mesoscopic integration method connecting microscopic and macroscopic topological features [5]. Since we can simulate individual levels of simulations according to current measurement data, we have to consider a modular architecture for multi-scopic simulations. In order to connect multi-scopic simulations, we need topological or graphical structures for the essential analysis without using all data.

For big data problems, growing neural gas (GNG) can cope well with hidden topological logic features [6–14]. However, we can apply the obtained topological structure to inference, search, and prediction based on graph theory and network science [3, 4]. Therefore, in this paper, we discuss the applicability of the obtained topological structure in macroscopic tracking. First, we apply multi-scale batch learning GNG (MS-BL-GNG) [6] to extract the topological structure for tracking. Next, we propose a prediction method by the adjacent matrix in the extracted topological structure. Next, we apply the proposed method to the topological tracking analysis in mobility support robots' navigation tasks. Finally, we discuss the effectiveness of the proposed method through several simulation results.

This paper is organized as follows. Section 2 explains multi-scopic simulations for human-robot interactions in the topological twin. Section 3 explains a method of MS-BL-GNG and macroscopic tracking analysis. Section 4 shows several numerical simulation results and discusses the effectiveness of the proposed method. Finally, Section 5 summarizes the paper, discusses the essence of the proposed approach, and gives the future direction of this study.

2 Multi-scopic Simulations for Topological Twin

2.1 Multi-scopic Approach

The method of multi-scopic simulations is important to deal with and integrate different timescale of simulation results in real time. Microscopic simulations have the highest resolution and the shortest time interval, allowing us to deal with the dynamics inside the object and the state of the inner space in detail. In order to realize the human-robot physical interactions, we will use a method of estimating human muscle state by using inverse dynamics based on neuromusculoskeletal simulations.

Mesoscopic simulations are designed to deal with the approximate rigid body dynamics between several objects in the human and robotic surroundings. A personal space can be considered as a typical mesoscopic simulation object. Especially, we focus on the ternary relation between humans, robots, and objects based on the perception and action in a surrounding environment. Intentional behavior is done based on the coupling of the perceptual system and action system under the constraint-based on social knowledge and rules through multi-agent AI simulations from the macroscopic level. Behavior is described by task-dependent rule sets.

Macroscopic simulations are used to deal with spatiotemporal relationships between objects without using dynamic simulations of public or social spaces. A task is described as a sequence of behaviors, and a path is described as a sequence of nodes in a topological map. Task domain knowledge is represented by graph neural networks and knowledge graphs.

Several similar concepts related to multi-scopic approaches have been discussed so far, e.g., multi-scale approaches and multi-resolution approaches. While multi-scopic approaches basically deal with different levels of physics in individual levels, multi-scale or multi-resolution approaches deal with the other scale or data resolution to reduce computational costs in the same physical or algorithmic method.

2.2 *Macroscopic Analysis Based on Human-Robot Mesoscopic Simulations*

This paper focuses on the macroscopic analysis based on a mesoscopic simulation of mobility support robots [5]. We assume P people and R robots in the mesoscopic simulation. We use two types of mobility support robots (MSR) – electric wheelchairs and robot-assisted walkers – where a robot is controlled by independent two independent wheel drives (Fig. 2a). By using fuzzy control based on Gaussian membership functions, all instances in the system (including people and robots) can take multi-objective behaviors of collision avoidance and target tracing [14–16].

The inputs to the fuzzy controller for collision avoidance and target tracing are the distance to the obstacle measured by the laser range finder (LRF) and the relative direction to a target point, respectively. A target point is given sequentially according to a human persona model. The collision avoidance and target tracing of a person are also controlled by multi-objective behavior coordination. While a person can stop when encountering a crowded situation, an MSR can go back in such a situation. The position data of humans and robots measured in the mesoscopic simulation (Fig. 2b) are transferred to the macroscopic simulation (Fig. 2c) through simple TCP/IP communication. We can use the trajectory data for the macroscopic analysis collected in the macroscopic simulation. Furthermore, we can control the sampling interval to obtain the required position data from the mesoscopic simulation according to the analysis result in the macroscopic simulation.

3 Topological Tracking

3.1 *Multi-scale Batch Learning Growing Neural Gas*

This section explains a method of MS-BL-GNG [6] with multi-scale data ranging from small to full size to achieve efficient and fast training of GNG. Stochastic gradient descent, mini-batch gradient descent, or batch gradient descent is used as a training approach. The original GNG uses an iterative method of stochastic gradient descent. Mini-batch gradient descent calculates the weight updates according to two or more data (smaller than the total size of the data set). The mini-batch gradient descent method is popularly used in training deep learning methods. Figure 3 is the update strategy for multi-scale batch learning (MS-BL) when the number of data is D and the maximum number of nodes is N.

In this case, the total size of the data set is divided into eight mini-batch sizes, and the amount of data at the phase level (L) is expressed as λ_L. The update of the reference vectors is done sequentially using individual mini-batch size data sets. After one mini-batch training update, a new node is added to the network. As a result, at lower phase levels, the number of nodes increases faster. Assuming that the number of nodes has reached the parameter (μ_L), the learning phase level is shifted

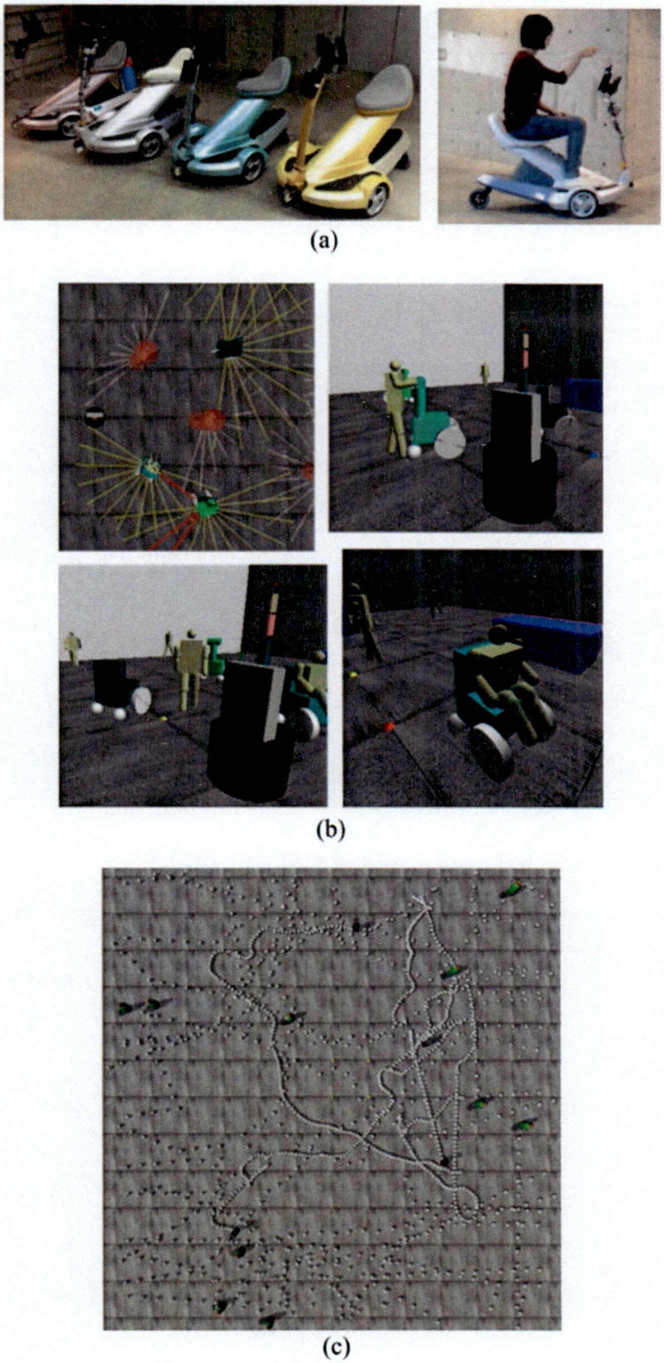

Fig. 2 Multi-scopic simulations on mobility support robots. (**a**) Example of intelligent senior cars; (**b**) mesoscopic simulation of mobility support robots (MSR) and humans; (**c**) macroscopic simulation for the tracking of humans and mobility support robots

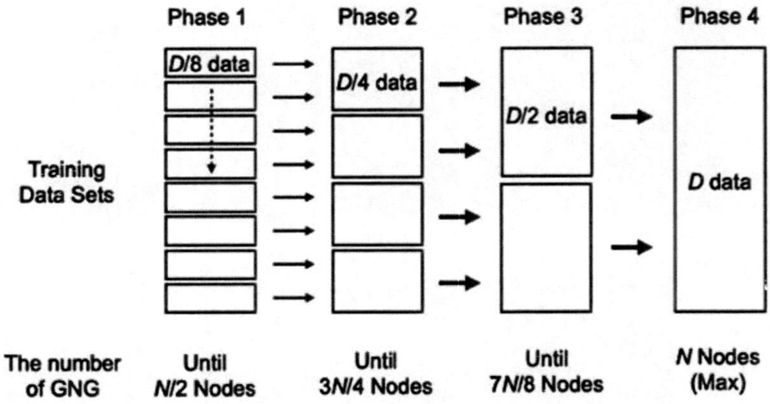

Fig. 3 Update of learning phase in MS-BL-GNG ($\lambda_L = \{D/8, D/4, D/2, D\}$, $\mu_L = \{N/2, 3\,N/4, 7\,N/8, N\}$ in the above example, D is the number of data and N is the maximum number of nodes)

to the next level. MS-BL-GNG initially tries to cover all the data sets roughly and finally is able to do fine topology mapping using all the data sets.

We explain the learning process of MS-BL-GNG. The notations used in GNG are shown as follows:

w_i: The nth dimensional vector of i-th node
A: A current set of nodes
E_i: Accumulated error variable
C_i: A set of nodes connected to the i-th node
$c_{i,j}$: The edge between the i-th and j-th nodes
$a_{i,j}$: Age of edge between the i-th and j-th nodes

Step 1: Generate two or three nodes and connect them with edges. Initialize the phase level ($L = 1$).
Step 2: Initialize the temporal edge connectivity ($c'_{i,j} = 0, \quad i, j \in A$), selection times ($x_i = 0, \ i \in A$), and the temporal weigh update ($\Delta w_i = 0, \ i \in A$). Start the multi-scale training with the iteration ($it = 1$).
Step 3: Update the temporal weigh update and the temporal edge connection according to a sample data

3.1. Select the nearest node (winner), s_1, and the second nearest node, s_2, with a sample data v which is selected as input to the network according to the probability of $p(v)$:

$$s_1 = \arg\min_{i \in A} \|v - w_i\| \tag{1}$$

$$s_2 = \arg\min_{i \in A \backslash \{s_1\}} \|v - w_i\| \tag{2}$$

3.2. Add up the errors between the input and reference vectors to update the accumulated errors:

$$E_{s_1} \leftarrow E_{s_1} + \left\| v - w_{s_1} \right\| \tag{3}$$

3.3. Calculate weight updates for the winner node and the nodes connected to it:

$$\Delta w_{s_1} \leftarrow \Delta w_{s_1} + \eta_1 \left(v - w_{s_1} \right) \tag{4}$$

$$\Delta w_j \leftarrow \Delta w_j + \eta_2 \left(v - w_j \right) \quad if \ \ c_{s_1,j} = 1 \tag{5}$$

Increment the selection times $\left(x_{s_1} + +, \quad x_j + + \ \ \left(j \in C_{s_1}\right)\right)$.
3.4. Update the temporal edge connectivity ($c'_{s_1,s_2} = 1$).

Step 4: Decrease the error variables of all nodes:

$$E_i \leftarrow E_i - \beta E_i \quad (\forall i \in A) \tag{6}$$

Step 5: Increment iteration times. Continue with Step 3 if the iteration time is not an integer multiple of λ_L.

Step 6: Update the weights by the MS-BL, update the edge connectivity, and remove the nodes $x_i = 0$:

$$w_i \leftarrow w_i + \Delta w_i / x_i, \quad \text{if } x_i > 0 \tag{7}$$

$$c_{i,j} = \begin{cases} c'_{i,j} & \text{if } c'_{i,j} = 1 \\ 0 & \text{otherwise} \end{cases} \tag{8}$$

Step 7: Insert a new node. When the number of nodes (A) reaches the parameter (μ_L), the learning phase level is incremented.

7.1. Select the node q with the maximum accumulated error:

$$q = \arg\max_{i \in A} E_i \tag{9}$$

7.2. Select the node f that has the largest accumulated error among the nodes connected to q.

7.3. Add a new node r to the network and interpolate its reference vector between q and f:

$$w_r = 0.5 \cdot \left(w_q + w_f\right) \tag{10}$$

7.4. Insert the edges connecting the new node r with nodes q and f, removing the original edge between q and f.

7.5. Decrease the error variables of q and f by a fraction α:

$$E_q \leftarrow E_q - \alpha E_q \tag{11}$$

$$E_f \leftarrow E_f - \alpha E_f \tag{12}$$

7.6. Interpolate the error variable of r from q and f:

$$E_r = 0.5 \cdot \left(E_q + E_f\right) \tag{13}$$

Step 8: Continue with Step 2 if the stopping criterion (e.g., network size or some performance measure) is not yet fulfilled.

3.2 *Topological Analysis by Adjacent Matrix*

Graph theory is helpful to discuss the feature of a topological structure. An adjacent matrix **A** of a topological structure is defined by $c_{i,j} = \{0, 1\}$. Since we don't consider the direction in the topological structure, the adjacent matrix is symmetrical about the main diagonal ($c_{i,j} = c_{j,i}$). The number of walks of the length k from the i-th node to j-th node is represented by the (i,j)-th entry $c^k_{i,j}$ of $\mathbf{A}^k$ where $c^1_{i,j} = c_{i,j}$, and we can obtain $\mathbf{A}^{k+1}$ by the matrix multiplication:

$$c^{k+1}_{i,j} = \sum_{h=1}^{n} c^k_{i,h} \cdot c^1_{h,j} \tag{14}$$

where the number of nodes is n. Figure 4 shows a simple example of the number of walks.

The adjacent matrix A in Fig. 4 and is represented by:

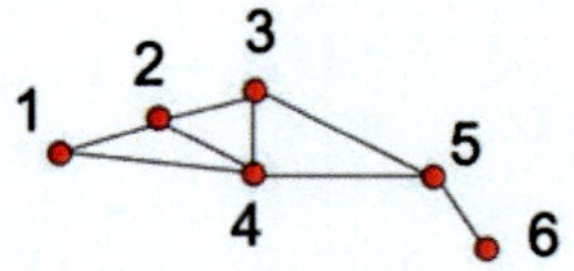

Fig. 4 An example of a graph ($n = 6$)

$$\mathbf{A} = \begin{pmatrix} 0\,1\,0\,1\,0\,0 \\ 1\,0\,1\,1\,0\,0 \\ 0\,1\,0\,1\,1\,0 \\ 1\,1\,1\,0\,1\,0 \\ 0\,0\,1\,1\,0\,1 \\ 0\,0\,0\,0\,1\,0 \end{pmatrix}, \mathbf{A}^2 = \begin{pmatrix} 2\,1\,2\,1\,1\,0 \\ 1\,3\,1\,2\,2\,0 \\ 2\,1\,3\,2\,1\,1 \\ 1\,2\,2\,4\,1\,1 \\ 1\,2\,1\,1\,3\,0 \\ 0\,0\,1\,1\,0\,1 \end{pmatrix} \tag{15}$$

Furthermore, the number of edges in this example is represented by the diagonal entries of $\mathbf{A}^2$. When an agent (a person or robot) is in the range of the first node, we can represent $\mathbf{h} = (1\ 0\ \cdots\ 0)^{\mathbf{T}}$ where the first node is selected and obtained $\mathbf{A} \cdot \mathbf{h} = \left(0\,1\,0\,1\,0\,0\right)^{\mathbf{T}}$ and $\mathbf{A}^2 \cdot \mathbf{h} = \left(2\,1\,2\,1\,1\,0\right)^{\mathbf{T}}$. If the i-th entry is positive, the agent can reach the i-th node. The value of the vector is calculated by the degree of variety of possible paths. Therefore, we use this idea for the tracking of persons, and mobility supports robots. The number of possible total paths from i-th to j-th node within k times of move is represented by:

$$\mathbf{S}_k = \sum_{i=1}^{k} \mathbf{A}^i. \tag{16}$$

Figure 5 shows an example of the tracking based on topological structure obtained by MS-BL-GNG: (a) a person and mobility support robot move to two target points repeatedly, (b) trajectories of their movements transferred from the mesoscopic simulation, (c) topological structure obtained by MS-BL-GNG where the number of maximal nodes in MS-BL-GNG is 40 and the number of training data is 800, and (d) adjacent nodes calculated by A^2 and the selected node s1 drawn by a large red sphere. However, the moving direction shown in Fig. 5d is not clear from the viewpoint of visualization.

In general, since we can approximately estimate the possible moving direction from the previous path, we use the following time series of tracking data of the b-th agent:

$$\mathbf{h}_b = \left(h_{b,1}\ h_{b,2}\ \cdots\ h_{b,n}\right)^{\mathbf{T}}, h_{b,i} \leftarrow \begin{cases} 1.0 & \text{if } i\text{th node is selected as } s_1 \text{ or } s_2 \\ \gamma h_{b,i} & \text{otherwise} \end{cases} \tag{17}$$

where γ is the temporal discount rate. By taking the negative possibility of the visit to the node visited previously into account, we represent the time series of tracking data in the following:

$$\mathbf{u}_b = \left(u_{b,1}\ u_{b,2}\ \cdots\ u_{b,n}\right)^{\mathbf{T}}, u_{b,i} \leftarrow \begin{cases} 1.0 & \text{if } i\text{th node is selected as } s_1 \text{ or } s_2 \\ -h_{b,i} & \text{otherwise} \end{cases} \tag{18}$$

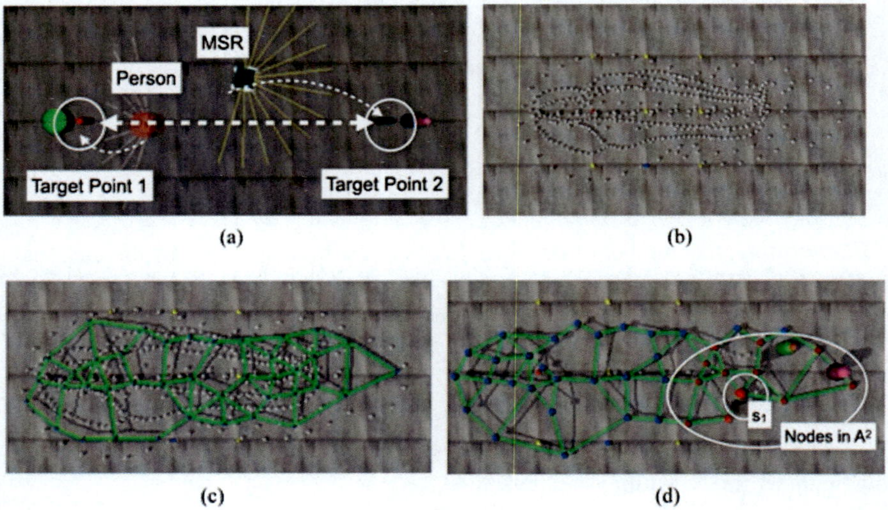

Fig. 5 An example of the tracking of MSR based on topological structure obtained by MS-BL-GNG. (**a**) A simple simulation example; (**b**) trajectories of movements; (**c**) an example of topological structure; (**d**) adjacent nodes in $\mathbf{A}^2$ (in red)

As a result, we can estimate the current situation in spatiotemporal tracking:

$$\mathbf{v}_b = \mathbf{S}_k \cdot \mathbf{u}_b \quad \text{where} \quad v_{b,i} \leftarrow \max\left(v_{b,i}, 0\right) \tag{19}$$

Figure 6 shows an example of the temporal tracking of MSR based on topological structure obtained by MS-BL-GNG: (a) adjacent nodes in A^2 (in red) and selected nodes (s1 and s2 drawn by a large red sphere) in the macroscopic simulation and (b) the current position of MSR and person in the mesoscopic simulation. We can understand the moving direction from Fig. 6 (a) and (c) by checking the relative positions of adjacent nodes in A^2. However, although the MSR approaches a person, such a situation is not reflected in the moving direction of the MSR in Fig. 6c.

Accordingly, we consider the positions of other agents in the moving direction of MSR in the following:

$$\tilde{\mathbf{u}} = \left(\tilde{u}_1\ \tilde{u}_2 \cdots \tilde{u}_n\right)^{\mathbf{T}}, \tilde{u}_i \leftarrow \begin{cases} 1 \text{ if } i\text{th node is selected as } s_1 \text{ or } s_2 \text{ of others} \\ 0 \qquad\qquad\qquad \text{otherwise} \end{cases} \tag{20}$$

As a result, we can estimate the current situation in spatiotemporal tracking based on the possibility of colliding with other agents:

$$\mathbf{v}_b \leftarrow \mathbf{v}_b - \tilde{\mathbf{v}}, \tilde{\mathbf{v}} = \mathbf{S}_k \cdot \tilde{\mathbf{u}} \tag{21}$$

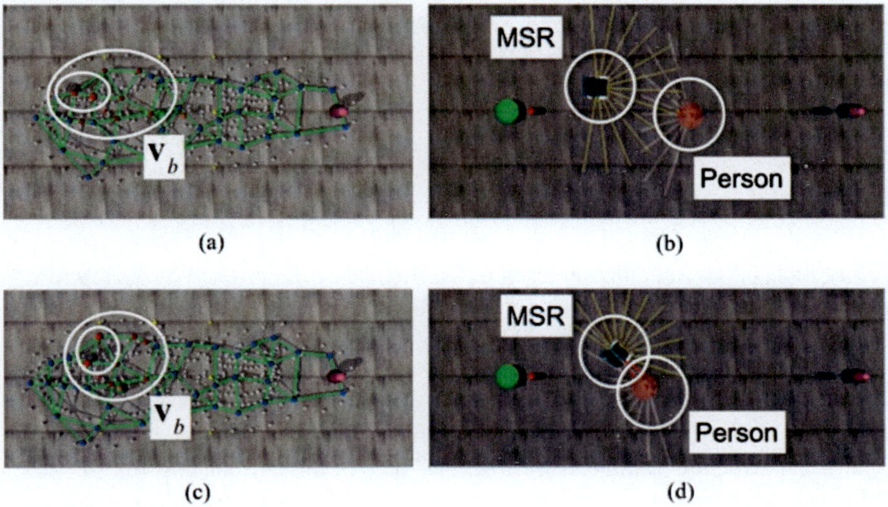

Fig. 6 An example of the temporal tracking of MSR based on topological structure obtained by MS-BL-GNG. (**a**) Adjacent nodes ($\mathbf{v}_b$) in $\mathbf{A}^2$ (in red) in Macro-S; (**b**) MSR and person in Meso-S; (**c**) adjacent nodes ($\mathbf{v}_b$) in $\mathbf{A}^2$ (in red) in Macro-S; (**d**) MSR and person in Meso-S

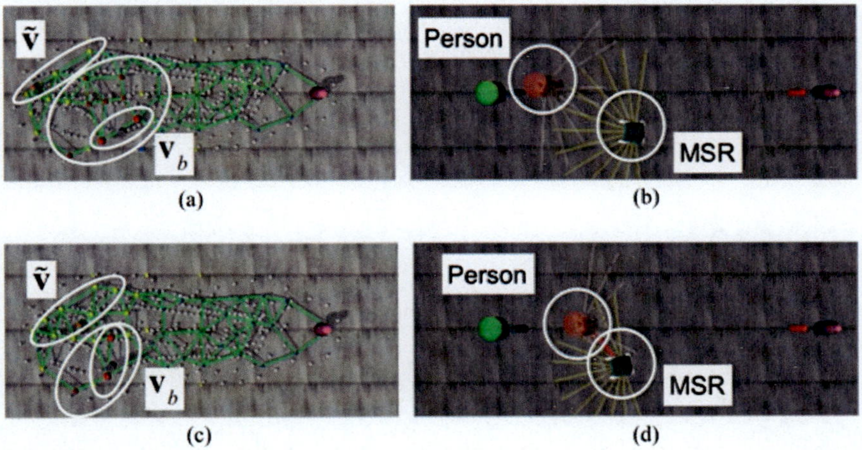

Fig. 7 An example of the temporal tracking of MSR with other agents based on topological structure obtained by MS-BL-GNG (**a**) Adjacent nodes ($\mathbf{v}_b$) in $\mathbf{A}^2$ (in red) in Macro-S; (**b**) MSR and person in Meso-S; (**c**) adjacent nodes ($\mathbf{v}_b$) in $\mathbf{A}^2$ (in red) in Macro-S; (**d**) MSR and person in Meso-S

Figure 7 shows an example of the temporal tracking of MSR with other agents based on topological structure obtained by MS-BL-GNG: (a) adjacent nodes in A^2 (in red) and selected nodes (s1 and s2 drawn by a large red sphere) and nodes with negative value (in yellow) in the macroscopic simulation and (b) the current position of MSR and person in the mesoscopic simulation. We can easily understand the current situation of MSR with other agents from Fig. 7a, c. Furthermore, the adjacent node in color shows the moving direction for collision avoidance with other agents in Fig. 7d.

4 Simulation Results

This section shows a simulation result of topological tracking based on the proposed method. The number of maximal nodes in MS-BL-GNG is 320, and the number of training data is 6400. The numbers of people and MRS are 4 and 2, respectively.

Figure 8 shows a simulation result of the temporal tracking of MSR with other agents based on topological structure obtained by MS-BL-GNG: (a) four persons and two mobility support robots move to one of four target points randomly, (b) topological structure obtained by MS-BL-GNG, (c) the current position of MSR and person in the mesoscopic simulation, and (d) adjacent nodes in A^2 (in red) and selected nodes (s_1 and s_2 drawn by a large red sphere) and nodes with negative value (in yellow) in the macroscopic simulation. We can easily understand the current situation of MSR with other agents from Fig. 8d, f. Figure 9 shows the change $\mathbf{V}_b$ in the temporal tracking of an MSR where the nodes are sorted according to the value of $\mathbf{V}_b$ in $\mathbf{A}^3$ in the temporal tracking of an MSR. The number of nodes in $\mathbf{A}^3$ is changing over time owing to the change of the adjacent nodes. The nodes with positive values and negative values can be used to calculate the attractive force and the repulsive force in an artificial potential field, respectively.

5 Summary

This paper proposed an analysis method of macroscopic topological tracking based on multi-scale batch learning growing neural gas. The topological mapping method is beneficial for extracting features from big data, and the graph theory is very useful for analyzing the topological features. First, we discussed how to use the adjacent matrix obtained by the connectivity of nodes in topological mapping to discuss the prediction in human and robot tracking. Next, we discussed how to update vectors corresponding to a moving trajectory of humans and robots. Finally, we showed the effectiveness of the proposed analysis method through several simulations results. The simulation results show that we can visualize the prediction in the tracking of humans and robots. The essence of using adjacent matric and vectors in the proposed method is in the simple mathematical calculation without the search in the topological map. Of course, we can find probable nodes where a robot may collide with humans in the future by using the graph search in the topological map.

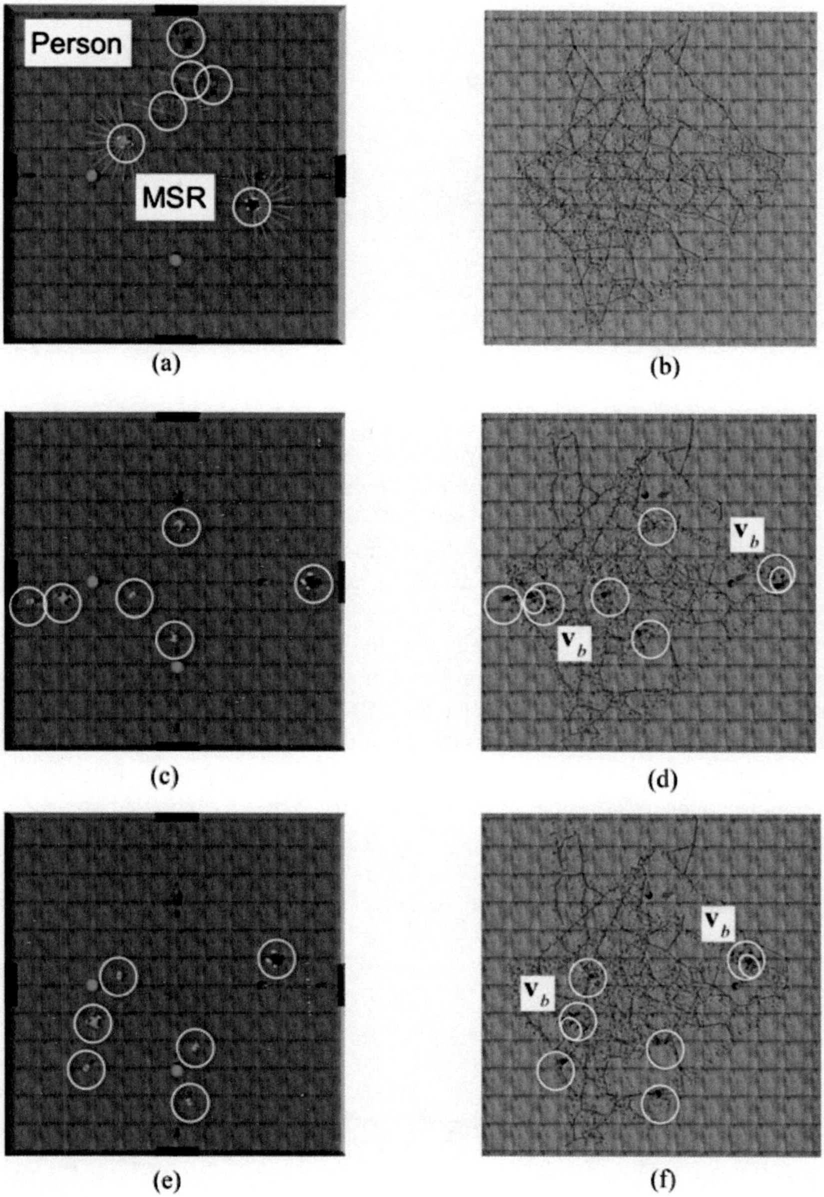

Fig. 8 An example of the temporal tracking of MSR with other agents based on topological structure obtained by MS-BL-GNG. (**a**) A simulation example in Meso-S; (**b**) an example of topological structure; (**c**) MSR and person in Meso-S; (**d**) adjacent nodes ($\mathbf{v}_b$) in $\mathbf{A}^2$ in Macro-S; (**e**) MSR and person in Meso-S; (**f**) adjacent nodes ($\mathbf{v}_b$) in $\mathbf{A}^2$ in Macro-S

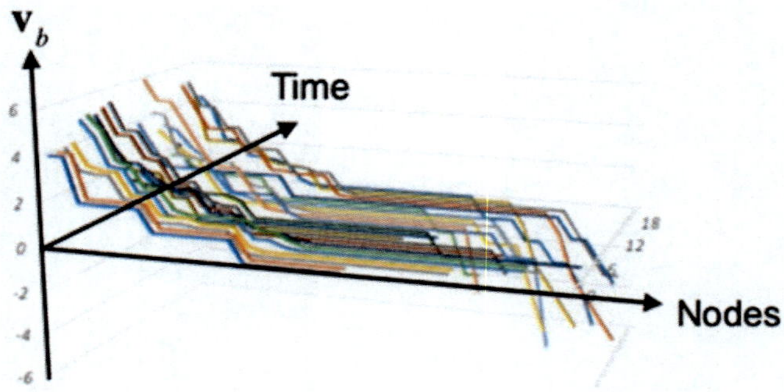

Fig. 9 The change of $\mathbf{v}_b$ in the temporal tracking of an MSR

Furthermore, a methodology on Markov chain Monte Carlo can be applied to predict the encountering probability of a robot with people. However, it takes much computational cost based on huge random samplings. MCMC cannot be practically applied to large data sets. On the other hand, we can obtain the approximate possibilities of future collisions by a simple mathematical calculation based on the proposed method. We will compare the computational cost and accuracy of the proposed method with stochastic simulations in the future.

In future work, we will apply the proposed method to temporal target generation in the macroscopic simulation to realize the collision avoidance in the mesoscopic simulation. In this way, we can discuss the meso-macro loop based on the bottom-up construction and top-down constrain in the multi-scopic simulations. Furthermore, we intend to connect the proposed method with the actual mobility support robots to show the effectiveness.

Acknowledgments This work was partially supported by JST [Moonshot RnD] [grant number JP-MJMS2034].

References

1. Wang, F.: The emergence of intelligent enterprises: from CPS to CPSS. IEEE Intell. Syst. **25**(04), 85–88 (2010)
2. Wang, P., Yang, L.T., Li, J., Chen, J., Hu, S.: Data fusion in cyber-physical-social systems: state-of-the-art and perspectives. Inf. Fusion. **51**, 42–57 (2019)
3. Bapat, R.B.: Graphs and Matrices. Hindustan Book Agency, New Delhi (2014)
4. Barabási, A.-L.: Network Science. Cambridge University Press, Cambridge (2016)
5. Oshio, S., Kaneko, K., Kubota, N.: Multi-scopic simulation for human-robot interactions based on multi-objective behavior coordination. The 7th International Workshop on Advanced Computational Intelligence and Intelligent Informatics (IWACIII 2021), 2021 (accepted)
6. Iwasa, M., Kubota, N., Toda, Y.: Multi-scale batch-learning growing neural gas for topological feature extraction in navigation of mobility support robots. The 7th International Workshop on Advanced Computational Intelligence and Intelligent Informatics (IWACIII 2021), 2021 (accepted)

7. Fritzke, B.: A self-organizing network that can follow non-stationary distributions. In: Artificial Neural Networks-ICANN'97, pp. 613–618. Springer, Berlin/Heidelberg (1997)
8. Fritzke, B.: A growing neural gas network learns topologies. Adv. Neural Inf. Proces. Syst. **7**, 625–632 (1995)
9. Martinetz, T.M., Schulten, K.J.: A "neural-gas" network learns topologies. Artif. Neural Netw. **1**, 397–402 (1991)
10. Fritzke, B.: Unsupervised clustering with growing cell structures. Neural Netw. **2**, 531–536 (1991)
11. Kohonen, T.: Self-Organizing Maps. Springer, Cham (2000)
12. Toda, Y., Matsuno, T., Minami, M.: Multilayer batch learning growing neural gas for learning multiscale topologies. J. Adv. Comput. Intell. Intell. Inform. **25**(6), 1011–1023 (2021) (in press)
13. Toda, Y., Chin, W., Kubota, N.: Unsupervised neural network based topological learning from point clouds for map building. 2017 International Symposium on Micro-NanoMechatronics and Human Science (MHS), 2017
14. Toda, Y., Kubota, N.: Topological structure learning based enclosing formation behavior for monitoring system. 2018 IEEE International Conference on Systems, Man, and Cybernetics (SMC), 2018, pp. 831–836
15. Fukuda, T., Kubota, N.: An intelligent robotic system based on a fuzzy approach. Proc. IEEE. **87**(9), 1448–1470 (1999)
16. Kubota, N., Aizawa, N.: Intelligent control of multi-agent system based on multi-objective behavior coordination. Proceedings of the 2008 IEEE International Conference on Fuzzy Systems (FUZZ-IEEE 2008), Hong Kong, China, June 1–6, 2008, pp. 1458–1463

Bibliographic Analysis of the Capacity and Applicability of Li-Fi Networks

Kelvin I. Seibt, Victor A. Kich, and Gabriel V. Heisler

1 Introduction

With the increase in the number of devices based on wireless communication due to the growth of the IoT (Internet of Things) area, the amount of consumed bandwidth is increasing in the same proportion. The most affected environments are public places where there are a huge number of users and may be a lot of obstacles, but depending on the case even our own home will have limitations that can be impost on the Wi-Fi network, like interference, attenuation. Such things can harm the user experience [7].

These problems arise from many fonts, but one of the principal factors that demand attention is the limit of the spectrum of the radio waves. Realizing that the spectrum of radio waves is saturated, and for consequence is expensive, we begin to think that technologies can come to solve or minimize these imposed limits. So, our research before initiated was a concern to find an environment with easy implementation and promissory results. In this way, we found the Li-Fi, an optic system of data communication.

Considering that all the workplaces require by law an appropriate illumination and that the electricity is a basic service in all world's countries, the Li-Fi shows to be suitable to mitigate these problems. The Li-Fi is a system that sends data through the visible and infrared (IR) light spectrum, and according to the first public demonstration of this technology, performed by Professor Harald Haas in the year

K. I. Seibt (✉)
University of Santa Cruz do Sul, Santa Cruz, Brazil

V. A. Kich · G. V. Heisler
Federal University of Santa Maria, Santa Maria, Brazil

D. Tang et al. (eds.), *Mobile Wireless Middleware, Operating Systems and Applications*, EAI/Springer Innovations in Communication and Computing,
https://doi.org/10.1007/978-3-030-98671-1_3

2011 in his lecture at TED Talks, an environment with a light-emitting diode (LED) lamp could provide a 10 Mbps transfer rate [3].

Having found this technology, this research has as objective to evaluate if the Li-Fi could act as a substitute for Wi-Fi or only as a complementary technology. It will be discussed areas that already have any applications to the technology, and the verdict will be made by listing comparisons with the Wi-Fi and exploring the deficiencies of this technology.

This evaluation will be done having as basis articles selected during the time frame from July 2011 to January 2020, considering the information of the book released in 2015 by the Professor Haas collaborative with Svilen Dimitrov, called "Principles of LED Light Communications: Towards Networked Li-Fi" [2].

2 Related Works

Three research bases were used for this search to quantify the commitment in this area that we are addressing. We used four terms in the searches, being them: Li-Fi, Wi-Fi, 5G, and efficiency. They were adopted filters to the pulished type of "Article" with the publication date from July of 2011 to January of 2020 and who contains in your content the terms cited above. The bases that were used was Coordenação de Aperfeiçoamento de Pessoal de Nível Superior (CAPES), Google Scholar, and the Institute of Electrical and Electronics Engineers (IEEE).

On a CAPES basis, when the term "Li-Fi" was searched individually, it generated 1986 results. By adding the term "Wi-Fi" to the search to get more works that would make a comparison between the technologies, this number has been reduced to 85. Then "5G" was added to have a return of more recent articles that were already making a connection with the future that has not yet been experienced, resulting in 65 results. Finally, the term "Efficiency" was added in an attempt to locate research that would quantify the results in a real environment. This is showed in Table 1.

The previous process was repeated in the Google Scholar and IEEE bases, presenting more promising results on Google Scholar, where it was assumed that the reason would be the broader search engine. See this in Tables 2 and 3.

As one perceives a very great influence of the creator of the technology, Haas, we find ourselves in the permission to search in other article bases, specifically by this author. Used also was his book released in 2015, called "Principles of LED Light Communications: Towards Networked Li-Fi" [2]. Therefore, the ScienceDirect

Table 1 Bibliometry applied to CAPES periodicals repository

Term	Li-Fi	Wi-Fi	5G	Performance
Li-Fi	1986	85	65	18
Wi-Fi	–	60,026	1413	754
5G	–	–	109,111	24,152
Performance	–	–	–	3,607,200

Table 2 Bibliometry applied to Google Scholar search tool

Term	Li-Fi	Wi-Fi	5G	Performance
Li-Fi	50,200	3010	814	512
Wi-Fi	–	880,000	38,300	25,600
5G	–	–	1,450,000	397,000
Performance	–	–	–	6,000,000

Table 3 Bibliometry applied to IEEE Explorer research base

Term	Li-Fi	Wi-Fi	5G	Performance
Li-Fi	262	42	8	0
Wi-Fi	–	53	2	0
5G	–	–	367	91
Performance	–	–	–	3147

research base has been added to our data set. In the following paragraphs, the articles and other references used are listed.

The article proposed by Khandal [7] is an analysis of the state of the technology before the launch of Professor Haas' book and follows a similar path to the one approached in this research. As registered in the year 2014, it offers a limited view of the progress made in this area, so it intended to update the information contained therein, using Haas' article [4] and his book [2]. After it, Bao et al. [1] solve the key technologies for realizing Li-Fi and present the state-of-the-art on each aspect. Posteriorly, Wu et al. [11] discussed the differences between homogeneous and heterogeneous networks regarding access point selection (APS), beyond proposing a two-stage APS method for hybrid Li-Fi/Wi-Fi networks.

The book was written by Haas and Dimitrov [2] and has played a key role in the construction of research since it has allowed building fundamental concepts for the initial understanding of the resources that are used in this technology. It also provided information on the evolution of the technology, since it can be compared with the article released by Professor Haas in 2018.

The article was published by Professor Haas in 2018 [4], where he addresses a more current view than the first article previously cited and also introduced why Li-Fi is considered a fifth-generation technology. It is intended to compare the results with those obtained by Khandal and Jain in 2014. Subsequently, Islim et al. [6] solve the suitable modulation techniques for Li-Fi including those that explore time, frequency, and color domains.

Finally, the last selected article was A Review Paper on Li-Fi Technology [9]. This article was produced in the year 2020, the year in which this research is being conducted, and therefore brings us analysis in the latest updates in this technology, therefore intending to analyze it against the work of Haas [4].

Given the above information in sets with the tabulated data, it has been concluded that the subject of this research has not yet been widely explored, and therefore, there is still a wide range of research that can be applied in the area.

3 Theoretical Background

To give the reader some background, we wanted to introduce some terms and mechanisms since these will be vital for the understanding of the discussions held throughout the work.

The sources of information about Li-Fi's communication and technology came from Haas' book [2] and conference [4]. The photodiode was presented with the help of data contained in the Hamamatsu datasheet [5] and the contents of the UFRGS classes [10].

3.1 Optical Communication

Optical communication is any form of telecommunication that uses light as the means of transmission. Originally called Optical Wireless Communication (OWC), it has evolved as a high-capacity technology, complementing the radio frequency communications. OWC systems use wavelength in the IR spectrum for IR communications and visible light spectrum in Visible Light Communication (VLC) [2].

3.2 Photodiode

It consists of an electronic device made of a semiconductor material (usually silicon). It has a semiconductor junction, which has the property of varying its electrical resistance according to the intensity of the light (the number of photons) in its incident [10]. The Hamamatsu Photonics company is a producer of this type of component and has models with specific applications and properties for various types of environment [5].

3.3 Li-Fi

Li-Fi is a VLC technology developed in 2010 through research conducted at the Edinburgh University and led by Professor Haas of Mobile Communications. It became popular to the public in 2011, when Haas presented the technology in his TED Talks, with the title "Wireless data from every light bulb" [3].

In 2011, the IEEE released a standard for VLC, called IEEE 802.15.7-2011, "IEEE Standard for Local and Metropolitan Area Networks, Part 15.7: Short-Range Wireless Optical Communication Using Visible Light" [2]. In Li-Fi, the data communication occurs basically by changing the light intensity of a LED, which in turn is modulated into a message signal and transmitted by the wireless optical channel and detected by a photodiode, which makes the demodulation and the recovery of the message clock [2].

4 Methodology

This study was applied through scientific research on a worldwide basis, applying a bibliometry according to the terms that were considered central to our theme. Thus, a foundation will be made in ideas and assumptions taken from conferences, articles, and books that present importance in the construction of the concepts of this work.

The observation method used was the conceptual-analytic, since concepts and ideas from other authors will be used, which are similar to our objectives, adding new content to them for the construction of scientific analysis on the object of study.

The comparison and discussion of the founded results will be carried out by means of explanatory research, giving more freedom to talk about an analysis that traverses several communication properties and allows them to assume more than one position on the subject during the analysis.

5 Results

According to Haas [2], the central idea of a Li-Fi wireless network is to complement heterogeneous wireless networks with radio, thus performing a relief for this spectrum and its amount of data traffic. In this way, the author already refers us to a partial response to the objective of this research, since it induces the reader that this technology will always need support, serving only as an "end" of the network.

In the following subtopics, we wanted to have a more specific view of what the limitations may make the Li-Fi a support tool for the network and not a complete structure.

5.1 Signal Modulation

A fully optical wireless network would require omnipresent coverage by optical peripherals. The likely candidates such as peripherals in VLC would be incoherent light LEDs because of their low cost. Due to the physical capacity of the components used, the information can only be encoded using light intensity. As a result, VLC can be used as an Intensity Modulation and Direct Detection (IM/DD) system, and this technique allows communication even if the lights appear visually switched off. The problem encountered with this modulation is that the signal must be real, unipolar, and not negative.

These premises eventually limit the consolidated modulation schemes used in radio wave communication. Techniques such as Pulse-Width Modulation (PWM), Pulse-Position Modulation (PPM), On–Off Keying (OOK), and Pulse-Amplitude Modulation (PAM) can be applied almost normally but, a modulation speeds increase, these schemes will suffer Inter-Symbol Interference (ISI) effects due to data frequency must be collected on the optical wireless network.

So a more resilient technique like Orthogonal Frequency-Division Multiplexing (OFDM) is needed, as it allows the use of adaptive bit and energy loading of different frequency sub-bands according to the properties of the communication channel. Besides, OFDMA is used in the 4th-generation (4G) Long-Term Evolution (LTE) communication standard for mobile phones [2]. Therefore, the application of OFDM in optical mobile networks would allow the use of the already established higher-level communication protocols used in IEEE 802.11 and LTE [2].

At this point, both the book and Haas' article remained with the same problem, but in 2018 two solutions are already proposed: enhanced unipolar OFDM (eU-OFDM) and Spectral and Energy Efficient OFDM (SEE OFDM). There were no relevant or divergent data reported on modulation in the articles by Khandal [7] and Jain and Singh [9].

5.2 Multi-User Access

A fully optical, seamless network solution can only be achieved with a suitable multiple access scheme that allows multiple users to share communication resources without any mutual cross-talk. The various access schemes used in RF communications can be adapted to the OWC, provided the necessary IM/DD-related modifications are made.

OFDM comes with a natural extension for multiple-access OFDMA. Single-carrier modulation schemes, such as PPM and PAM, require an additional multiple-access technique, such as Frequency Division Multiple Access (FDMA), Time Division Multiple Access (TDMA), or Code Division Multiple Access (CDMA) [2].

In the OWC, there is an alternative option to achieve multiple access, which is the color of the LED, and the corresponding technique for this is Wavelength Division Multiple Access (WDMA). This scheme can reduce the complexity of signal processing at the expense of increasing hardware complexity.

In this sense, no changes were reported in any of the articles, remaining in this format until the present moment. One of the factors that may harm the experience of multiple users on a Li-Fi topology will be treated in Sect. 5.6.

5.3 Network Structure

Ready-to-use technologies such as Power Line Communication (PLC) and Power over Ethernet (PoE) are viable back-haul solutions for Li-Fi installations [2]. In this sense, Wi-Fi ends up having a greater resilience, because it does not depend on a structure that uses data and energy in its communication media.

Li-Fi then ends up being limited to a range of structures so that it can operate in such a way that there is no restructuring of the network at the site. No changes were reported in any of the articles, remaining in this format until the present moment.

5.4 Upload

So far, research has focused mainly on maximizing transmission speeds on a single unidirectional link. However, for a complete Li-Fi communication system, full-duplex communication is required, i.e., an uplink connection from the mobile terminals to the optical access point (AP) must be provided. The existing duplex techniques used in RF such as Time Division Duplexing (TDD) and Frequency Division Duplexing (FDD) can be considered, where the downlink and uplink are separated by different time intervals or frequency ranges, respectively. However, the FDD is more difficult to perceive due to the limited bandwidth of front-end devices and because the super hetero-dynamic is not used in IM/DD systems.

The most appropriate duplexing technique in Li-Fi is Wavelength Division Duplexing (WDD), where the two communication channels are established by different electromagnetic wavelengths or using IR transmission as a viable option to establish an uplink communication channel [2].

A first commercially available full-duplex Li-Fi modem using IR light for the uplink channel was recently announced by pureLiFi [8]. The author still reminds us that a router that needed a visible light to send the data would also be something that would generate distraction [4].

This becomes relevant as there is an imbalance in traffic in favor of downlink in current wireless communication systems.

5.5 Topology

In Li-Fi, the topology adopted is point-to-point, while Wi-Fi is multi-point [9].

While Singh [9] says the technology needs a line of sight, Haas [2] complements that a point-to-point redirection can be done to make communication possible in a certain way, not targeted. Techniques for a really non-target vision are still being studied and also depend on the surface of reflection.

5.6 Cells

In the past, wireless cellular communication has benefited significantly from reducing the distance between stations of cellular base stations. By reducing cell size, network spectral efficiency has increased by two orders of magnitude over the past 25 years.

More recently, different cell layers composed of microcells, picocells, and femtocells have been introduced. These networks are called heterogeneous networks. Femtocells are short-range base stations, low transmission power, low cost, and plug-and-play that are targeted for internal deployment to improve coverage. They use cable Internet or Digital Subscriber Broadband (DSL) to return to the operator's main network. The deployment of femtocells increases frequency reuse and, therefore, the transfer rate per unit area within the system, as they usually share the same bandwidth with the macrocellular network.

However, uncoordinated and random deployment of small cells also causes additional inter- and intracell interference, which imposes a limit on how dense these small base stations can be deployed before the interference begins to compensate for all the frequency reuse gains.

The concept of small cells, however, can easily be extended to VLC to overcome the high interference generated by near-spectrum radiofrequency reuse in heterogeneous networks. Optical AP is referred to as an attoc cell. Since it operates in the optical spectrum, the optical cell does not interfere with the macrocellular network. The optic cell not only improves the internal coverage, but as it does not generate any additional interference, it is capable of enhancing the capacity of radio frequency wireless networks [2].

The range of cells in the 5G of technology will probably be up to 25 meters [4]. This is the information that comes according to the idea of Haas in 2015 when he said that the intention was that these Li-Fi cells have a maximum range of 5 meters, so there is no overlapping of lights.

5.7 Data Rates

The data rate achieved by technology depends directly on the optical technology employed. Currently, most commercialized LEDs are composed of a high brightness blue LED with a phosphor cover, which converts this blue light into a yellow tone. When these two colors combine, they form the white light. This is the most cost-effective way to produce the equipment, but this conversion, which decreases the frequency of response and high frequencies, is greatly attenuated. Consequently, the bandwidth of this type of LED is around 2 Mhz [4].

However, by ignoring the cost and looking for a higher shipping fee, we can achieve incredible values. By following a speed scale, adding a blue filter to the receiver in the scheme reported in the previous paragraph, it is possible to reach metrics of 1 Gbps.

With red, green, and blue (RGB) LEDs, up to 8 Gbps is achieved, as the light is naturally generated through the mixing of colors and not through chemical components. Finally, the most promising structure is laser-based lights, which have

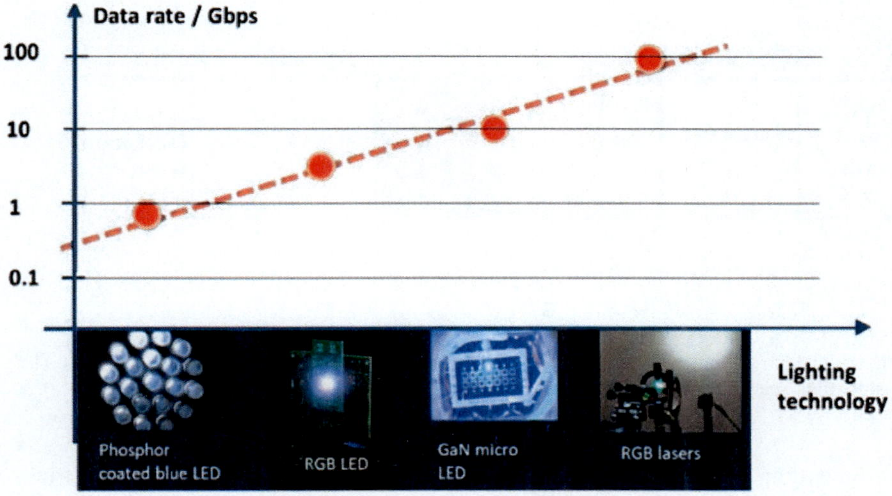

Fig. 1 Graph of the technologies that can be employed with VLC and its data rates, from [4]

shown that a range of 100 Gbps is achievable according to [4]. This is shown in Fig. 1.

5.8 Use Cases

The cases of use presented by Khandal and Jain [7], as well as Haas in his book [2], are closed office or home environments, i.e., short distances. However, in his latest article [4], he already brings us a look, focused also on safety, pointing out the use in intrinsically safe environments, such as petrochemical plants and oil platforms where radio frequency is often forbidden. Singh finally brings us applications in the underwater environment, which is convenient with the idea of platforms in the middle of the sea, from Haas [4]. The light can then pass through the saltwater and works in dense regions as a sensor [9].

5.9 Li-Fi and 5G

Currently, most technologies use RF waves for data transmission, but the growing demand for data traffic requires increasing bandwidth, which is not supported by RF. The visible light and IR specter are approximately 2,600 times the size of the entire 300 GHz radio frequency spectrum, appearing as a viable alternative to meet the demand for data traffic. This is possible to see in Fig. 2.

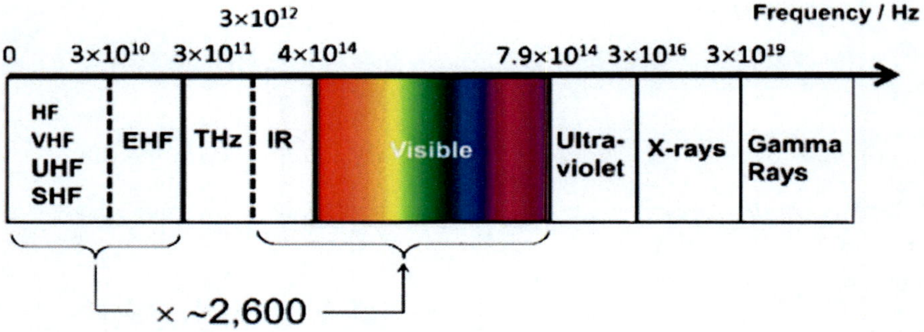

Fig. 2 The RF spectrum is only a fraction of the entire electromagnetic spectrum, from [4]. The visible light spectrum and the IR spectrum are unregulated and offer 780 THz bandwidth

According to Haas [4], Li-Fi is a technology that will impact many industries. With the high data rate that was arranged in the subtopic "Data Rate" and the bandwidth discussed in the paragraph above, Li-Fi fits as a 5G technology. It also adds that the adoption of the OWC standard, together with the use of RGB lasers, can bring a new revolution to the area of communication.

There would then be a large-scale unlocking of IoT in industry 4.0 applications, as well as the growth of projects that are currently just being born. An example of Light as a Service (LaaS) in the power industry, merging two major industries: the wireless communication industry and the power industry. In the near future, the LED light lamp will serve thousands of applications and will also be an integral part of smart cities, autonomous cars, smart homes, and IoT.

However, this technology is currently under development and needs more resilience in other environments where a quality connection is required, in addition to overcoming the barriers imposed by the optical medium itself and the technology employed.

6 Conclusion

As Haas approached at the beginning of his book, Li-Fi is still just a Wi-Fi supporting technology and still has a long way to go to replace it. The complexity added to hardware to accomplish a good multi-user access and data rates make it a barrier to entry for ordinary consumers.

The point-to-point topology points that multiply spaces in the same build will need a previous preparation of network scheme. Without that, it becomes necessary for the acquisition of multiply devices to make possible the redirect of the signal. This brings up high use of hardware that, with the necessity of evolution of methodologies applied to Li-Fi, can quickly become obsolete, bringing an accumulation of electronic waste.

However, it can already be observed that it is, in fact, a 5G technology with high transmission and bandwidth rates.

Based on the selected articles, the technology has not yet undergone many changes since the regulation and creation of a protocol for VLC communication. The same problems are still perceived, but some research results to add more solutions to these problems are already visible.

Based on the results and tests cited by Haas, there is a great bet that the market will receive more prototypes and companies focused on this market, such as pureLiFi. The existing technology today is still very affected by the interference between its communication peripherals and the noise by the particularity of sunlight impulses.

With the consolidation of this technology, you will have the possibility of using in places that are impossible to use Wi-Fi. Safety is a beneficial side, where the use of light to transmit data because it does not penetrate through walls. On highways for traffic control applications, such as cars, they can have LED lights and can communicate with each other and predict accidents.

Therefore, the key point for the search for solutions to problems in Li-Fi and your implementation seem to forward automatically with the passage of time, the lack of radio band. Haas already made alarming predictions of increased data use in 2015, and today it is a reality. We have some devices that demand communication with other devices in the environment, make queries in the cloud, and still have advanced artificial intelligence algorithms. This is a perfect environment for Li-Fi to act and its potential to be explored.

After reading the book, it is noted that there are still a large number of topics to be addressed within this technology. What stood out in terms of the need for study are the areas of signal modulation and access to multi-users. Therefore, themes directed to the discussion or solution of these deficiencies would have much relevance in the area of communication.

References

1. Bao, X., Yu, G., Dai, J., Zhu, X.: Li-Fi: Light fidelity-A survey. Wirel. Netw. **21**(6), 1879–1889 (2015)
2. Dimitrov, S., Haas, H.: Principles of LED Light Communications: Towards Networked Li-Fi, 1st edn. Cambridge University, Cambridge (2015)
3. Haas, H.: Wireless data from every light bulb. In: TEDGlobal 2011. TED, New York (2011)
4. Haas, H.: Li-Fi is a paradigm-shifting 5G technology. In: Reviews in Physics. Elsevier, Amsterdam (2018)
5. Hamamatsu Photonics, K.: Datasheet: Si pin photodiode s6801/s6968 series (2020). https://www.hamamatsu.com/resources/pdf/ssd/s6801_etc_kpin1-046e.pdf. Accessed on 2020-06-13
6. Islim, M.S., Haas, H.: Modulation techniques for Li-Fi. ZTE Communications **14**(2), 29–40 (2019)
7. Khandal, D., Jain, S.: Li-Fi (light fidelity): The future technology in wireless communication. In: International Journal of Information and Computation Technology, vol. 4, pp. 1–7 (2014)

8. pureLiFi: Home—pureLiFi—connectivity is evolving. https://pureli-fi.com/ (2020). accessed on 2020-06-14
9. Singh, R.: A review paper on Li-Fi (light fidelity) technology. In: EasyChair Preprint, vol. 1, pp. 1–9. EasyChair Preprint, New York (2020)
10. UFRGS: Sensores-fotodiodos. http://www.if.ufrgs.br/mpef/mef004/200-61/Cesar/SENSORES-Fotodiodo.html (2013). Accessed on 2020-06-13
11. Wu, X., Safari, M., Haas, H.: Access point selection for hybrid Li-Fi and Wi-Fi networks. IEEE Trans. Commun. **65**(12), 5375–5385 (2017)

Layered-MAC: An Energy-Protected and Efficient Protocol for Wireless Sensor Networks

Ekereuke Udoh and Vladimir Getov

1 Introduction

This paper documents the development of a new MAC layer protocol which demonstrates an ability to tackle denial-of-sleep (DoSL) attacks better than the existing duty-cycled protocols. Battery-powered sensors usually have a network lifetime of 3.5 years [2]. However, a successful DoSL attack can reduce the life span of these sensors to 3 days [3–6]. Such significant loss of energy requires a deeper look into the problem, hence the need for a protocol that is energy-efficient and protects against these attacks. Our protocol is implemented on two different platforms: a simulated environment using OMNET++ [7] and a small proof-of-concept prototype using physical devices such as Sun SPOT [8]. More emphasis is placed on the simulation results rather than the real experiments using physical devices. This is because the simulation platform gives room for scalability analysis allowing a variety of bridge sizes and numbers of nodes, whereas the physical platform is limited to just three devices. Hence, the physical devices are used as a proof-of-concept to support the simulation results. The developed protocol also includes some inherent security features as part of the process of tackling DoSL attacks [9, 10].

Our solution is evaluated based on how energy loss it eliminates as well as how it responds in the event of a DoSL attack. These sources of energy loss include overhearing, idle listening, control packets overhead, and collisions [11]. The new

This is a modified and extended version of previously published work [1].

E. Udoh (✉) · V. Getov
Distributed and Intelligent Systems Research Group, University of Westminster, London, UK
e-mail: ekereuke.udoh@qa.com; v.s.getov@westminster.ac.uk

D. Tang et al. (eds.), *Mobile Wireless Middleware, Operating Systems and Applications*, EAI/Springer Innovations in Communication and Computing,
https://doi.org/10.1007/978-3-030-98671-1_4

protocol tackles each of these sources of energy loss in a unique and secure way. The research begins by identifying the requirements of the protocol, specifying these requirements, and prioritizing them using a technique called MoSCoW which indicates four priority categories – (a) must have; (b) should have; (c) could have; and (d) would have [12]. Furthermore, different designs of the semantics of the protocol are produced and discussed. Algorithms are then produced for the protocol. These algorithms are implemented on the OMNET++ simulator and on a small testbed with the Sun SPOT sensor devices. The language used for the simulation and device implementation is discussed critically based on different criteria. The paper also provides an evaluation of the Layered-MAC protocol in comparison to several existing MAC layer protocols.

The rest of this paper is organized as follows. Section 2 provides a summary of related work elsewhere. Section 3 lays emphasis on the software engineering aspects such as requirements gathering and functional and non-functional requirements of the protocol. Section 4 shows the results of the simulations including simulations under DoSL attack. Section 5 concludes the paper and discusses future work.

2 Related Work

It is pertinent to note that in the context of DoSL, several approaches exist to curb these attacks. However, most of them do not take energy efficiency into consideration, and even when they do, throughput becomes a trade-off which could become counterproductive in the long run. The most notable existing approaches include Gateway-MAC (GMAC) [13], hash-based Scheme [14], clustered adaptive rate limiting [15], fake schedule switch Scheme [16], absorbing Markov chain (AMC) model [17], secure wakeup Scheme [18, 19], zero-knowledge protocol [20], and cross-layer mechanism [21].

One of the existing protocols that have geared towards energy efficiency as well as security is the GMAC protocol. GMAC protocol uses the idea of a central management where nodes are divided into clusters and each cluster has a gateway node. One of the strategies used in tackling DoSL attacks is by understanding the impact of a failed node on the entire network lifetime. This is evidenced in [22] where the most critical node is assessed in terms of the impact of its elimination on the network lifetime. On the other hand, in [15, 23], an intrusion detection scheme (IDS) is proposed whereby a DoS attacks are detected before it has any impact thereby making it preventive. In [24], focus is placed on creating hard-to-guess tokens/beacons which prevents attackers from easily guessing tokens that are aimed at depleting battery life. In [25], a cluster-based security protocol which uses digital signatures is proposed; however, this does not consider energy efficiency. Another protocol based on public-key cryptography is proposed in [26]. However, this protocol seems to introduce a lot of overhead that comes with key exchange and management.

A generic framework that optimizes the performance of existing clustering protocols such as UHEED by using simulated annealing and K-Beam algorithms is proposed in [27]. However, this is mainly aimed at clustering and routing protocols. In [7], the relationship between node density and certain network parameters such as the received signal strength indicator (RSSI) and the link quality indicator (LQI) is analyzed, with reference to DoSL attacks.

With regard to clustering, one protocol which considers energy efficiency is GMAC [28] and is also intended to guard against broadcast attacks which target the MAC layer. GMAC uses a cluster-based approach, thereby enhancing security; however, the clustering limits the network architecture thereby making it have a low level of autonomy with regard to network architecture. The clustered adaptive rate limiting approach uses a host-based intrusion detection system to limit the rate of activity by the radio as a way of conserving energy and curbing against the broadcast attack with a downside if reducing throughput [29].

Furthermore, the hash-based scheme also uses clustering in addition to hashing but focuses on reducing the overhead involved in selecting a cluster head better than random vote scheme and round robin Scheme [30].

Fake schedule switch Scheme [31] as the name implies creates a false schedule and uses an offensive approach rather than a defensive approach by sending the wrong schedule when an acknowledgment is not received, but one limitation is that it only applies to protocols that support acknowledgment-based communication.

The secure wakeup Scheme [32] appears to have a mechanism where the packet can be inspected while the node is still in a low-power sleep state. This is achieved by the radio being able to hold a list of tokens and carry out an authentication. It would have been good to highlight how much energy is spent.

The absorbing Markov chain [33] works by attempting to predict the expected death time of a sensor and using that as a benchmark to detect a DoSL attack by monitoring the network traffic. The limitation of this technique is that it is detective in nature and not corrective or preventive.

Anomaly detection is the technique used by hierarchical collaborative model [23]. It attempts to achieve this not with one node but multiple nodes thereby balancing out the load. The downside is that achieving this requires a lot of packet overhead.

Zero-knowledge protocol [34] uses RSA key generation, hash generation, and distribution and interlock protocol to achieve security, however with little or no focus on the energy costs of the technique. It does help against man-in-the-middle and replay attacks but no evidence to curbing against DoSL attacks.

In [35], to curb DoSL attacks, a combination of firefly algorithm, Hopfield neural network, and RSA is applied in additional to the considering mobility of the sink node. The mobility of the sink node is based on the premise that in fixed-sink networks, nodes that are closer to the sink tend to drain energy faster as they act as a proxy for the rest of the nodes, and that mobility of the sink node will solve this problem.

In [36], k-nearest neighbor classification algorithm is implemented using Python libraries such as scikit-learn, NumPy, and Pandas. This machine learning algorithm uses data from incoming traffic to a node to detect a DoSL attack. Four traffic-related features are used to train the model in addition to heuristically determined rules, and the results of the confusion matrix show an 87% accuracy.

In [24], the vulnerabilities present in MAC layer protocols such as SMAC, TMAC, and B-MAC are highlighted. For SMAC, an attacker can use false SYN messages with longer time than the transmission frames. For B-MAC, an attacker can take advantage of the preambles. For TMAC, an attacker can take advantage of the adaptive timeouts.

3 Requirements Analysis

First, performance tuning was done on existing protocols to understand the impact of certain parameters like duty cycling and beacon interval fraction and transmit power on metrics like energy consumption and number of transmitted packets. The performance tuning was done using a protocol called TunableMAC which was created using two languages – NED and C++ – and runs on OMNET++ framework as part of the Castalia simulator. NED was used to define the network including its parameters and gates, while C++ was used to define the behavior of the MAC protocol. The platform for these languages is OMNET++, and this was used alongside a framework for wireless sensor networks called Castalia. The C++ codes consisted of two files – a header file which contained a declaration of the variables and methods and another file which contained an initialization of the variables and implementation of the methods. The reverse engineering was done to understand the sequence and effect of the methods as well as the states of the variables. Hence a sequence diagram and state diagram are produced for the TunableMAC protocol. The diagrams provide a better understanding of where to insert the algorithms for the new protocol.

In building the new protocol, a traditional software development life cycle (SDLC) was used, particularly the **incremental model/iterative model**. This involved building the protocol in small increments. Each increment involved all the stages of the SDLC which are described briefly below:

- Requirements gathering/analysis. This stage involves understanding the problem and deciding on what needs to be done. In some cases, these two stages are split individually, but considering the scope of this protocol, not many requirements are required. Hence the two stages can be combined into one. The requirement is then clearly specified.
- Design. This stage involves producing a blueprint of the internal workings of the system be it high-level or low-level design. One design could be a flow chart showing the flow of information in the protocol. Another design could be a sequence diagram showing the sequence of method calls for the new protocol. A class diagram showing the methods and variables of the new protocol is also an important design to include.

- Implementation. At this stage, the coding will be done either for the simulator in C++ and NED or for the Sun SPOT sensor in Java [37]. This stage involves testing the codes to first check that they meet the requirements and that they perform better than existing protocols at tackling DoSL attacks.

3.1 Requirements Identification

Problem Statement DoSL attacks can have a strong negative impact on the life span of battery-powered wireless sensors. Considering that the radio is the major source of energy loss, these attacks take advantage of the MAC layer, which is responsible for access to the radio, and use certain techniques to prevent the radio from sleeping thereby reducing the life span of the sensor. While there have been proposed solutions and techniques to tackling these attacks, only one of these solutions (GMAC) has been incorporated into a protocol and tested on a real device. There is therefore a need for more MAC layer protocols that have a form of security against DoSL attacks while aiming at maintaining the same or similar level of throughput and latency as protocols that do not have these security measures.

Protocol Requirements Only two levels of headings should be numbered. The protocol should be able to detect a DoSL attack and take measures to reduce its impact. In a case where the protocol is not able to detect the DoSL attack on time, it should take measures to reduce the other sources of energy loss that are not because of an attack. In this way the sensor can have enough energy to continue functioning until it detects the attack. To detect the attack, the first step is to understand the possible attack strategies that could be used:

- Attack from an unauthorized authenticated node: In this scenario, the node's identity is verified and valid; however, the action of the node is not authorized.
- Attack from an authorized and authenticated node: This is a more dangerous scenario as it is more difficult to detect such a node. In this case the entire identity has been compromised. Sybil node attacks fall under this category.
- Attack from an unauthenticated and unauthorized node: This is the least dangerous of the three strategies.

Then, it is important to identify the target because an attack on a sink node would have more impact than an attack on a cluster head. Similarly, an attack on a cluster head would have more impact than an attack on a normal node. After identifying the target, the next step is to get some data about the attacker node beginning with its address and RSSI and LQI for that node. After the node has been identified, the next step is to isolate the attacker and make the network inaccessible by that node.

The life cycle of the MAC layer is divided into four stages as follows:

- The start-up stage involves initializing the variables with information about the packets, sensors, and communication as well as getting the node to sleep if there

is no information from the radio layer or there is nothing left in the buffer to send to the network layer. At this stage, the cluster heads will also be set up.
- The transmit stage involves transmitting information received from the network layer to the radio layer or transmitting information received from the radio layer to the network layer.
- The carrier sensing stage is before transmitting, when a node may want to apply some CSMA techniques or use request-to-send (RTS) or clear-to-send (CTS) packets to avoid collisions and overhearing. While RTS/CTS could be helpful in avoiding collisions, it has one disadvantage of increasing the control packet overhead which further increases the energy consumption. CSMA on the other had has some back-off techniques that work based on probability and may not always be accurate and could lead to deadlock problems where a node is not able to transmit because if waiting endlessly for an opportunity to transmit.
- The receive stage involves staying in a receive mode and waiting for information from the radio layer which is coming from another node. The data received must be checked to know the type of data (control packet or actual data).

3.2 Functional Requirements

As mentioned earlier, the MoSCoW technique is used to prioritize the requirements based on the following four categories:

Must Have. The sink node should be able to get the RSSI and LQI values of every sensor it receives data from. Each node should know how far it is from the sink node and use that to decide who becomes a cluster head. The protocol should be able to adjust the duty cycle at runtime based on the traffic. The protocol should allow cluster heads to be appointed and rotated at intervals if need be. The protocol should allow for a node to be isolated from the network when it has been discovered to be an attacker node.

Should Have. Nodes should be able find the least expensive route to communicate their data. Cluster heads should be able to communicate using code division multiple access.

Could Have. Supervised learning could be applied on the data collected from the base station.

Will Not Have. Nodes cannot be powered by solar energy.

3.3 Algorithm Design

Algorithm for the Layered-MAC Protocol The algorithm flow chart depicted in Fig. 1 highlights the steps involved in setting up the MAC layer before

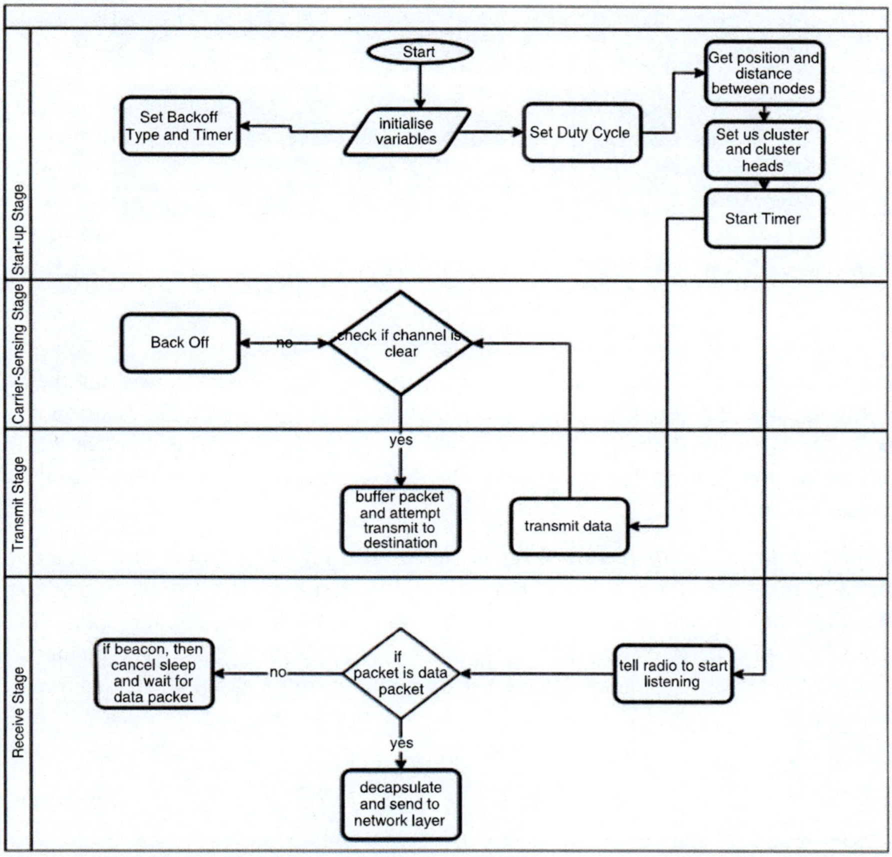

Fig. 1 Flow chart showing the Layered-MAC communication

the communication process starts. More specifically, the main steps include the following actions:

- MAC layer receives the number of nodes from the application layer.
- Sink node gets the distance of all nodes.
- Sink node appoints the node with closest proximity as a cluster head.
- If a cluster head has more than five nodes assigned to it, then another cluster head is appointed.
- Nodes must only communicate to their cluster heads not to other nodes.
- The cluster head then passes the information to the sink nodes. If the sink node is too far from the cluster head, then the data is passed to other cluster heads closer to the sink node.
- After every 5 min, a new cluster head is appointed to ensure security and to also manage the energy efficiency.

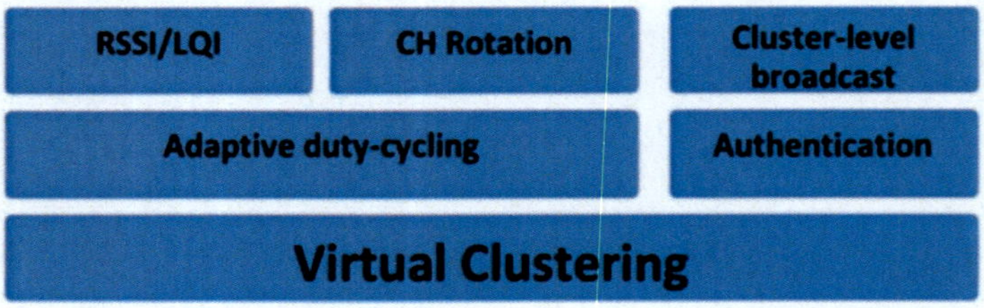

Fig. 2 Component architecture model of the Layered-MAC protocol

Figure 2 shows the conceptual design of the Layered-MAC protocol. The lowest layer is virtual clustering which involves grouping the sensors based on their proximity. The benefits of keeping the clusters virtual are that if the position of the sensor is changed, then the cluster can be reconfigured [38]. The adaptive duty cycling is adopted form TMAC whereby the duty cycle automatically adjusts to the amount of traffic.

Algorithm for Position and Distance of Nodes Getting the positions and distance between nodes involves using a points-based system/GPS to get the x and y coordinates of the sensor. After getting the x and y coordinates for each node, the next step would be to calculate the distance between the two nodes. The distance between the nodes is calculated using the following formula based on Pythagorean theorem:

$$d = \sqrt{(x_2 - x_1)^2 + (y_2 - y_1)^2} \tag{1}$$

where d is the distance; x_2 is the x coordinate for node 2; x_1 is the x coordinate for node 1; y_2 is the y coordinate for node 2; and y_1 is the y coordinate for node 1. However, this method may not work for sensors located in areas where the GPS does not work. Another way to achieve this is by using the Castalia framework to get the RSSI and the LQI.

The algorithm for position and distance of nodes involves the following steps:

- Each node waits for a random time and makes a broadcast.
- The broadcast packet contains the schedule, and each node follows the schedule it receives.
- Each node also keeps the RSSI/LQI of the packet it receives.
- Get position of nodes as described above.
- Base station creates a map of the distance between nodes using RSSI.
- Identify best distance from each node.
- Create virtual clusters based on best distance.
- Use CDMA to communicate between cluster heads.
- Stop.

Table 1 shows a test plan based on the sources of energy loss in wireless sensors.

Table 1 Test cases and corresponding actions

ID	Test case	Description	Actions
1	Collision	This checks that the protocol plays a role in reducing reduction	Compare the number of transmitted and received packets
2	Control overhead	This checks that the protocol reduces the control overhead	Track the size of data used for control overhead
3	Idle listening	This checks that the protocol plays a significant role in reducing	Measure how long a node stays idle before transmitting
4	Overhearing	This checks if the protocol reduces the chances of a node hearing a packet meant for another node	Measure how much energy is wasted listening to packets meant for other nodes

Creating a map of the distance between the nodes by the base station involves creating a matrix that maps the distance between each node. For example, if there are ten nodes, each node maps the distance with nine other nodes. The mapping is based on the RSSI and LQI values of the sensor nodes. Based on the map, the best distance for each node is then calculated. The reason for using distance is to enhance the energy efficiency in the nodes when transmitting data. Using the best distance, the clusters are then created with cluster heads managing nodes within the closest distance. Only the cluster heads communicate directly with the base station/sink node. The cluster heads will be changed at intervals to increase security. Thus, the algorithm for cluster creation involves the following three steps:

- If nodes have the same schedule, they belong to the same cluster.
- Cluster nodes can only communicate with their cluster head.
- Cluster heads then communicate with the sink node.

Finally, CDMA is used only for communication between cluster heads to ensure security and prevent DoSL attacks. This stage has more to do with the physical layer.

4 Protocol Simulation

4.1 Test Plan

4.2 Experimental Results

This subsection presents the OMNET++ simulator results. Furthermore, the results for the Layered-MAC protocol are then compared with the results from SMAC and TMAC.

Consumed Energy The first graph in Fig. 3a shows the energy consumed for Layered-MAC under different duty cycles in a 40 m bridge. The consumed energy increases as the duty cycle increases. The second graph in Fig. 3b shows the energy

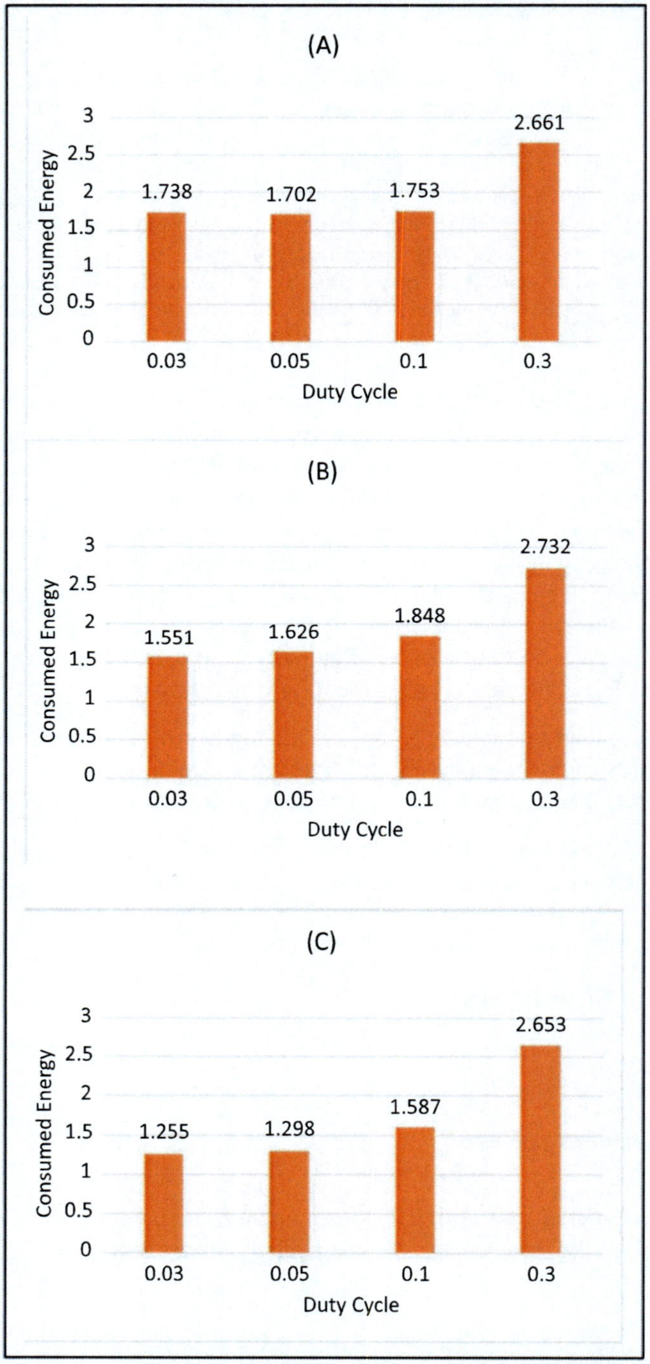

Fig. 3 Energy consumption for Layered-MAC on 40 m, 200 m, and 1000 m bridge. (**a**) Energy consumed for layered-MAC for 40 m bridge; (**b**) energy consumed for layered-MAC for 200 m bridge; (**c**) Energy consumed for layered-MAC for 1000 m bridge

consumption for Layered-MAC in a 200 m bridge. The energy consumed in the 200 m appears less that the energy consumed in 40 m bridge. The third graph in Fig. 3c shows the energy consumption for the 1000 m bridge for which energy consumption is the lowest.

Energy Comparisons The graphs above show a comparison of the Layered-MAC with two other protocols – TMAC and SMAC. The comparison is based on the energy consumption of the sensors.

The graph in Fig. 4a shows that TMAC consumes the least energy while Layered-MAC consumes the most energy. However, this does not take into consideration the packet reception at the sink.

In the second graph, energy is compared on a 200 m bridge for SMAC, TMAC, and Layered-MAC, and Layered-MAC consumes the most energy at 1.848, while TMAC consumes the least energy at 1.358.

Reception Comparisons The second graph in Fig. 5b shows that the Layered-MAC has a better reception at the sink in terms of data throughput with success of 11.088 and failure of 17.088 compared to SMAC and TMC which are significantly lower. The third graph (Fig. 5c) shows that the Layered-MAC has a better reception at the sink in terms of data throughput. There is a greater amount of succeeded and failed packets than in TMAC and SMAC.

In conclusion, while the energy consumption in total is higher for the Layered-MAC than TMAC and SMAC, the reception results show that the Layered-MAC provides better performance and even energy consumption when measured in terms of energy per successfully received packets.

Another interesting observation is that although more energy is spent as the bridge size increases, the successfully received packets show a downward trend for both TMAC and SMAC which is slightly different for the Layered-MAC.

Hence, on a 40 m bridge for Layered-MAC, 1.752 is spent for 21.05 received packets. On a 200 m bridge, 1.848 is spent for 11.088 received packets and 1.587 for 2.223 received packets on a 1000 m bridge. Summing up the received packets for TMAC and SMAC put together still does not get up to half the reception for Layered-MAC.

4.3 Simulation Results for the Protocols Under Attack

Simulation Scenario Only two levels of headings should be numbered. The Layered-MAC protocol is used alongside two other protocols, SMAC and TMAC, in the simulation to understand the energy consumption and reception under an attack.

For this scenario, a 200 m bridge is used with three of the nodes as compromised malicious nodes which use a broadcast attack to stop nodes from sleeping. The broadcast attack is carried out by continuously flooding the network with broadcast messages from these three nodes. The simulation is about the structural health

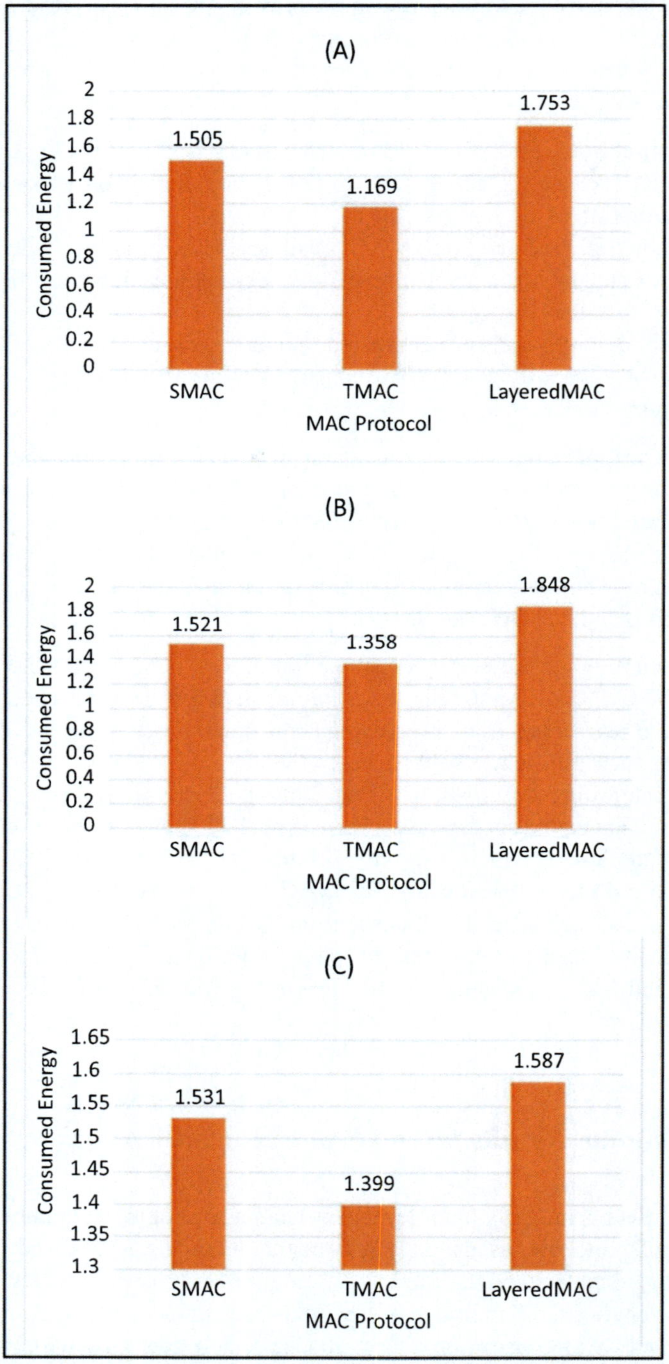

Fig. 4 Comparison for SMAC, TMAC, and Layered-MAC on 40 m, 200 m, and 1000 m bridge. (**a**) SMAC, TMAC and layered-MAC for 40 m bridge; (**b**) SMAC, TMAC and layered-MAC for 200 m bridge; (**c**) SMAC, TMAC and layered-MAC for 1000 m bridge

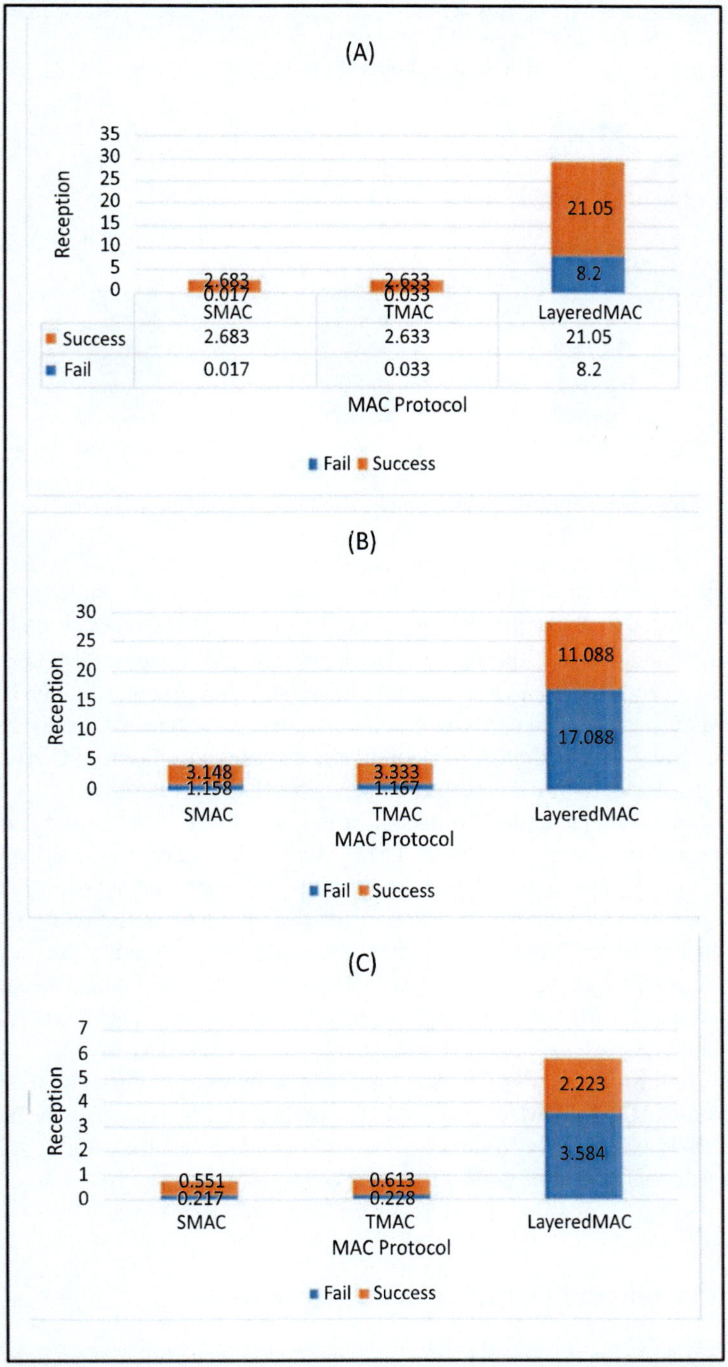

Fig. 5 Reception comparison for SMAC, TMAC, and Layered-MAC on 40 m, 200 m, and 1000 m bridge. (a) Reception for SMAC, TMAC and layered-MAC for 40 m bridge; (a) Reception for SMAC, TMAC and layered-MAC for 200 m bridge; (a) Reception for SMAC, TMAC and layered-MAC for 1000 m bridge

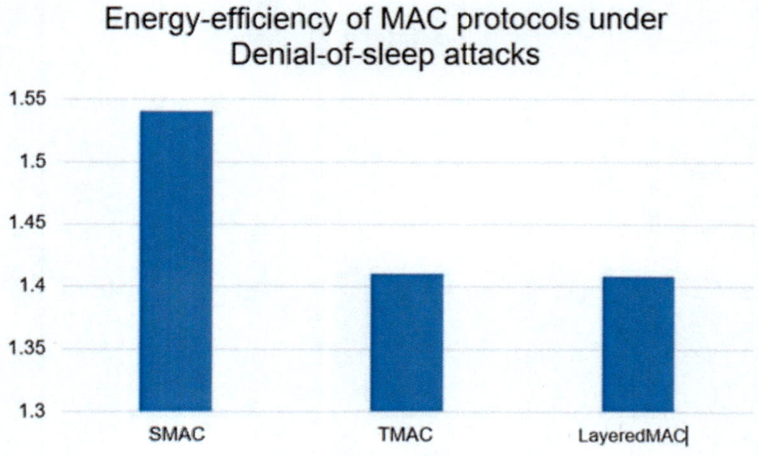

Fig. 6 Energy efficiency of MAC protocols under DoSL attack simulation

monitoring of a bridge. Sensing nodes are placed in a grid with a sink node in the middle. A car moves on the bridge every 5 minutes and triggers nodes along its path.

Figures 6 and 7 show the energy efficiency and throughput for three MAC protocols, respectively, including the new Layered-MAC protocol, under a DoSL attack. GMAC has not been used in the comparisons for two reasons: firstly, because there is no model of it in the Castalia simulator and secondly because GMAC does not give any considerations to reception (throughput) at the sink.

Figure 6 shows that under the DoSL broadcast attack, SMAC consumes the highest amount of energy at 1.54 J. TMAC consumes much lower energy than SMAC at 1.41 J. The Layered-MAC consumes the least energy slightly below TMAC at 1.408 J. The fixed duty cycling for SMAC justifies the relatively high energy consumption. TMAC on the other hand supports adaptive duty cycling; hence, there is better energy efficiency than SMAC. The Layered-MAC on the other hand goes a step further than just adaptive duty cycling to also detect signal strength and link quality, hence the slightly better energy-saving than TMAC.

In terms of throughput, Fig. 7 shows significant difference in throughput between the Layered-MAC and TMAC and SMAC combined. This is partly because of the Layered-MAC's ability to detect a malicious node and adjust duty cycling to bypass/isolate the malicious node.

5 Conclusion and Future Work

It is important to look at how the research questions discussed in the introduction have been addressed. CSMA with collision avoidance is used to prevent collisions when two nodes are trying to communicate at the same time. Acknowledgments

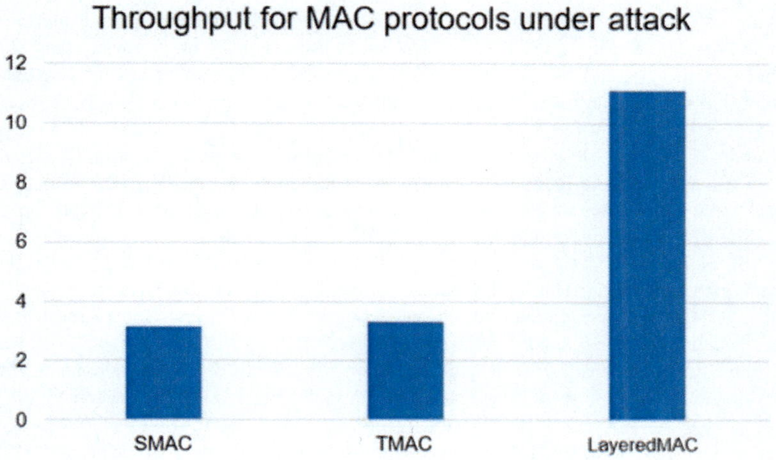

Fig. 7 Throughput of MAC protocols under DoSL attack

are not required to reduce the number of control packets. Minimizing the number of broadcasts by applying distance measurements and using a routing-by-rumor approach helps reduce overhearing. Idle listening is reduced by adaptive duty cycling. One of the benefits of this new protocol is that it is multilayered and touches on different aspects. First it deals with virtual clustering [27], authentication, RSSI, and LQI measurements which help with security. It also measures the distance between nodes and supports adaptive duty cycling which helps with energy consumption. Furthermore, our Layered-MAC protocol is tested alongside two other protocols under DoSL attacks, and the performance as well as the energy efficiency is significantly better.

One of the areas for future work is to investigate how machine learning techniques can be applied to data collected by the sensor nodes to further enhance energy efficiency and security against DoSL attacks. The algorithms will be run on the machine connected to the base station, and then the output from the learning is then passed across to the nodes as an update.

Acknowledgments This work was partially supported via a doctoral research scholarship grant by the University of Westminster.

References

1. Udoh, E.: An energy aware and secure mac protocol for tackling denial of sleep attacks in wireless sensor networks. PhD thesis, University of Westminster (2019)
2. Raymond, D.R., Midkiff, S.F.: Denial-of-service in wireless sensor networks: attacks and defenses.IEEE Pervasive Comput. **7**(1), 74–81, January–March IEEE (2008). https://doi.org/10.1109/MPRV.2008.6

3. Shakhov, V., Koo, I.: Depletion-of-battery attack: specificity, modelling and analysis. In: Special Issue on Security in IoT Enabled Sensors, Sensors, vol. 18, no. 6, Switzerland (2018)
4. Gelenbe, E., Kadioglu, Y.M.: Battery attacks on sensors. In: Proceedings of the International Symposium on Computer and Information Sciences, Security Workshop, pp. 1–10, Springer (2018)
5. Gehrmann, C., Tiloca, M., Hoglund, R.: SMACK: short message authentication check against battery exhaustion in the Internet of Things. In: Proceedings of the 2015 12th Annual IEEE International Conference on Sensing, Communication, and Networking (SECON), pp. 274–282, IEEE (2015) https://doi.org/10.1109/SAHCN.2015.7338326
6. Rani, S., Naidu, S.K.: Mitigation of energy depletion in wireless ad-hoc sensor networks through path optimization. Int. J. Comput. Netw. Appl. **2**(1), 1–11 (2015)
7. Boulis, A.: Castalia A Simulator for Wireless Sensor Networks and Body Area Networks User's Manual, Version 3.2. NICTA, Sydney (2011)
8. Horveliur, C.M.: Sun SPOTs: a great solution for small device development. https://www.oracle.com/technetwork/server-storage/ts-4868-1-159029.pdf. Last accessed 20 Nov 2018
9. Udoh, E., Getov, V.: Performance and energy-tuning methodology for wireless sensor networks using TunableMAC. In: 2020 International Conference on Communications, Computing, Cybersecurity, and Informatics (CCCI), pp. 1–5. IEEE, Sharja, UAE (2020). https://doi.org/10.1109/CCCI49893.2020.9256744
10. Udoh, E., Getov, V.: Proactive energy-efficiency: evaluation of duty-cycled mac protocols in wireless sensor networks. In: 2018 International Conference on Computer, Information and Telecommunication Systems (CITS), pp. 1–5. IEEE, Colmar, France (2018). https://doi.org/10.1109/CITS.2018.8440194
11. Sinha, P., Jha, V.K., Rai, S.K., Bhushan, B.: Security vulnerabilities, attacks and countermeasures in wireless sensor networks at various layers of OSI reference model: a survey. In: Proceedings of IEEE International Conference on Signal Processing and Communication, ICSPC 2017, vols. 2018-January, no. July, pp. 288–293, IEEE, Coimbatore, India (2017)
12. Agile Business Consortium: The DSDM agile project framework. Available at: https://www.agilebusiness.org/content/moscow-prioritisation. Last accessed 20 Nov 2018 (2014)
13. Brownfield, M.I.: Energy-efficient wireless sensor network MAC protocol. PhD thesis, Virginia Polytechnic Institute and State University (2006)
14. Pirretti, M., Zhu, S., Vijaykrishnan, N., Mcdaniel, P., Kandemir, M., Brooks, R.: The sleep deprivation attack in sensor networks: analysis and methods of defense. Int. J. Distrib. Sens. Netw. **2**(3), 267–287. SAGE (2006)
15. Vaseer, G., Ghai, G., Patheja, P.S.: A novel intrusion detection algorithm: an AODV routing protocol case study. In: Proceedings of the 2017 IEEE International Symposium on Nanoelectronic and Information Systems, *iNIS 2017*, vols. 2018-February, pp. 111–116, IEEE, Bhopal, India (2018)
16. Chen, C., Hui, L., Pei, Q., Ning, L., Qingquan, P.: An effective scheme for defending denial-of-sleep attack in wireless sensor networks. In: 5th International Conference on Information Assurance and Security, IAS. IEEE Xi'an China (2009)
17. Bhattasali, T., Chaki, R.: AMC model for denial of sleep attack detection. J. Recent Res. Trends. Cornell University arXiv Prepr. arXiv1203.1777 (2012)
18. Falk, R., Hof, H.J.: Fighting insomnia: a secure wake-up scheme for wireless sensor networks. In: Proceedings of the Third International Conference on Emerging Security Information, Systems and Technologies, pp. 191–196 (2009). https://doi.org/10.1109/SECURWARE.2009.36
19. Montoya, M., Bacles-Min, S., Molnos, A., Fournier, J.J.: SWARD: a secure wakeup RaDio against denial-of-service on IoT devices. In: 11th ACM Conference on Security and Privacy in Wireless and Mobile Networks (WiSec'18). CCSD, Stockholm, Sweden (2018). https://doi.org/10.1145/3212480.3212488.cea-01922847
20. Naik, S., Shekokar, N.: Conservation of Energy in Wireless Sensor Network by Preventing Denial of Sleep Attack. In: Procedia Computer Science. Elsevier, Amsterdam (2015)

21. Hsueh, C.T., Wen, C.Y., Ouyang, Y.C.: A secure scheme against power exhausting attacks in hierarchical wireless sensor networks. IEEE Sens. J. **15**(6), 3590–3602. IEEE (2015)
22. Yuksel, A., Uzun, E., Tavli, B.: The impact of elimination of the most critical node on wireless sensor network lifetime. In: IEEE Sensors Applications Symposium, Proceedings, pp. 1–5, IEEE, Zadar, Croatia (2015)
23. Wang, J., Jiang, S., Fapojuwo, A.O.: A protocol layer trust-based intrusion detection scheme for wireless sensor networks. In: Mauri, J., Han, G. (eds.) Smart Communication Protocols and Algorithms for Sensor Networks, vol. 17, no. 6, Sensors, Switzerland (2017)
24. Islam, M.N.U., Fahmin, A., Hossain, M.S., et al.: Denial-of-service attacks on wireless sensor network and defense techniques. Wirel. Pers. Commun. **116**, 1993–2021 (2021). https://doi.org/10.1007/s11277-020-07776-3
25. Ferng, H.W., Khoa, N.M.: On security of wireless sensor networks: a data authentication protocol using digital signature. Wirel. Netw. **23**(4), 1113–1131. Springer (2017)
26. Capossele, A.T., Cervo, V., Petrioli, C., Spenza, D.: Counteracting denial-of-sleep attacks in wake-up-radio-based sensing systems. In: Proceedings of 2016 13th Annual IEEE International Conference on Sensing, Communication, and Networking (SECON), 2016, pp. 1–9, IEEE, London, UK (2016). https://doi.org/10.1109/SAHCN.2016.7732978
27. Udoh, E., Getov, V.: Performance analysis of denial-of-sleep attack-prone MAC protocols in wireless sensor networks. In: Proceedings of 2018 UKSim-AMSS 20th International Conference on Computer Modelling and Simulation, pp. 151–156. IEEE, Cambridge (2018). https://doi.org/10.1109/UKSim.2018.00038
28. Krentzn, K.F., Graupner, H., Meinel, C.: Countering three denial-of-sleep attacks on Contiki-MAC. In: Proceedings of 2017 International Conference on Embedded Wireless Systems and Networks, pp. 108–119, Junction Publishing US (2017)
29. Uher, J., Mennecke, R.G., Farroha, B.S.: Denial of sleep attacks in Bluetooth low energy wireless sensor networks. In: Proceedings of the IEEE Military Communications Conference MILCOM, pp. 1231–1236, IEEE Baltimore, MD, USA (2016)
30. Qiu, L., Jiang, W., Zhang, W, Li, P.: Wireless injection attacks based on fake data injection in TinyOS. In: Proceedings – International Symposium on Parallel Architectures, Algorithms and Programming, vols. 2016-Januaary, pp. 236–242, PAAP, Nainjing, China (2016)
31. Krentz, K.F., Graupner, H.: Denial-of-sleep-resilient session key establishment for IEEE 802.15.4 security: from adaptive to responsive. In: Proceedings of 2018 International Conference on Embedded Wireless Systems and Networks, pp. 25–36, EWSN, Madrid, Spain (2018)
32. Pawar, P., Nielsen, R.H., Prasad, N.R., Prasad, R.: GSHMAC: green and secure hybrid medium access control for wireless sensor network. Wirel. Pers. Commun. **100**(2), 267–281 (2018)
33. Krentz, K.F., Meinel, C.: Denial-of-sleep defenses for IEEE 802.15.4 coordinated sampled listening (CSL). Comput. Netw. **148**, 60–71 (2019). https://doi.org/10.1016/j.comnet.2018.10.021
34. Osanaiye, O.A., Alfa, A.S., Hancke, G.P.: Denial of service defence for resource availability in wireless sensor networks. IEEE Access. **6**, 6975–7004. IEEE (2018)
35. Fotohi, R., Firoozi Bari, S.: A novel countermeasure technique to protect WSN against denial-of-sleep attacks using firefly and Hopfield neural network (HNN) algorithms. J. Supercomput. **76**, 6860–6886 (2020). https://doi.org/10.1007/s11227-019-03131-x
36. Desnitsky, V.: Approach to machine learning based attack detection in wireless sensor networks. 2020 International Russian Automation Conference (RusAutoCon), pp. 767–771, IEEE Sochi, Russia, (2020).https://doi.org/10.1109/RusAutoCon49822.2020.9208085
37. Oracle Labs: Sun SPOT Programmers Manual. Available at: https://bit.ly/2DlFFiy. Last accessed 1 Jan 2019 (2011)
38. Isaiadis, S., Getov, V.: Integrating mobile devices into the grid: design considerations and evaluation. In: Proceedings of Euro-Par 2005 Conference, LNCS, vol. 3648, pp. 1080–1088. Springer (2005)

Study on Urban Travel Volume During the Outbreak of COVID-19

Fang Xie, Zengping Zhang, Baojun Sun, Yinghao Zhou, Bo Li, and Yu Han

1 Introduction

Urban travel intensity is a reflection of the function of urban structure. The long-term interaction between crowd activity and spatial structure forms the unique function and structure of a city [1, 2]. The travel intensity is affected by time and space structure, and travel volumes vary from time to time and region to region. Understanding population distribution and changes in urban space is important for urban planning and government decision-making. Understanding the population distribution and mobility in urban space can provide relevant suggestions for the government to prevent and control new crown disease, especially in the context of the current global epidemic that has not been effectively contained.

The analysis of the distribution and change of urban traffic flow is usually carried out from two dimensions of space and time [3–5]. The spatial dimension reflects the regional structure, including autocorrelation and heterogeneity. The time dimension reflects the fluctuation of urban travel volume with time. But the urban travel volume is not all stable changes; in a few cases, there will be abnormal fluctuations. Therefore, some scholars have added the space-time dimension to the model [4, 5] in recent years. The modeling of the space-time dimension is to detect the abnormal fluctuation of traffic flow. With the continuous updating of statistical information technology, regional traffic analysis has been gradually explored by scholars. In

F. Xie (✉) · Y. Han
School of Business Administration, Inner Mongolia University of Finance and Economics, Hohhot, China

Z. Zhang (✉) · B. Sun · Y. Zhou · B. Li
School of Computer Information and Management, Inner Mongolia University of Finance and Economics, Hohhot, China
e-mail: 100002170@imufe.edu.cn

D. Tang et al. (eds.), *Mobile Wireless Middleware, Operating Systems and Applications*, EAI/Springer Innovations in Communication and Computing,
https://doi.org/10.1007/978-3-030-98671-1_5

recent year, Bayesian hierarchical model is widely used to analyze the distribution and change of spatiotemporal data [4–6], which allows us to include both spatial and temporal effects, while the Bayesian hierarchical model can be simulated using Markov chain Monte Carlo (MCMC) method. With a large number of iterations of the method, accurate a posteriori values can be obtained.

Some researchers believe that mobile phone data can be aggregated over time to reflect population distribution [7, 8]. The intensity of city trip refers to the ratio of the city trip flow to the number of people living in the city. The intensity of intracity trip reflects the activity of a city's trip flow and may be influenced by the latent function of the city. Therefore, to reveal the spatial and temporal distribution characteristics of urban trip flow fluctuation will help to deepen the understanding of the general pattern of population fluctuation within a city especially the regions with large population fluctuation in a period. In this paper, we use Baidu open-source data-in-city travel intensity data in Hubei province and establish a Bayesian hierarchical model to explore the predictable spatial-temporal pattern of urban travel intensity; the travel intensity in the city is divided into three dimensions: time, space, and space. A stable posterior distribution is obtained through a large number of iterations of the model, and the spatial, temporal, and spatial structure of Hubei province is analyzed based on the posterior results.

2 Data and Methods

2.1 Study Area and Data

Area in our research is Hubei province, which is divided into 17 areas by city level. Each city is an analytical unit.

Our data are daily travel intensity data of 17 cities in Hubei province from 2020.01.01 to 2020.05.01. The data is the open source data of Baidu website. The dataset does not contain any personal information.

2.2 Hierarchical Spatiotemporal Model

Q_{it} is the travel intensity of area i, Day t. The travel intensity in city is influenced not only by the spatial dimension and the time dimension but also by the space-time dimension.

In general, the urban travel volume belongs to Poisson distribution. So, we establish a spatiotemporal model with a logarithmic connection in the Poisson framework. In this paper, the Poisson likelihood method of data is used to model the regional counting variability under the condition of unknown parameters. In the first level of the model, suppose the change in traffic Q_{it} follows a Poisson distribution with mean Q_{it} for each region:

$$Q_{it} \sim \text{Poisson}\,(Q_{it}) \tag{1}$$

In the second layer of the model, we decompose Q_{it} on logarithmic scale into intercept α, space effect ζ_i, time effect ω_t, and space-time effect ε_{it}:

$$\log\,(Q_{it}) = \alpha + \zeta_i + \omega_t + \varepsilon_{it} \tag{2}$$

The spatial effect ζ_i shows the spatial distribution of the relative travel intensity ratios in the study area. The time effect ξ_t represents the overall time trend of travel intensity in all regions. The spatiotemporal effect ε_{it} represents the predictable portion of the relative ratio. We treat these effects as random variables and assign a reasonable prior distribution according to the previous experience to reflect the spatial and temporal autocorrelation and heterogeneity in order to better capture the potential structure of risk [3, 4].

The conditional autoregressive (CAR) models are usually used to express the spatial correlation for the spatial effect ζ_i, and the spatial correlation is represented by the specified neighborhood graph, which defines the neighborhood set of each region i. Associated with this graph is an adjacency matrix of $N \times N$, represented by $W = (w_{ij})$. If i and j are adjacent, then $w_{ij} = 1$; otherwise $w_{ij} = 0$. Then, we use the convolution priori [9] to model the spatial random effect. Then, it could be written as:

$$\zeta_i \sim N\left(\mu_i, \sigma_\zeta^2\right)$$

$$\mu_i \sim \text{CAR}\left(W, \sigma_\mu^2\right) \tag{3}$$

Both conditional autoregressive model and convolution model can be used to model time effect. Therefore, time effect can be modeled as:

$$\omega_t \sim N\left(\gamma_t, \sigma_\omega^2\right)$$

$$\gamma_t \sim \text{CAR}\left(W, \sigma_\gamma^2\right) \tag{4}$$

For the spatiotemporal effect ε_{it}, it reflects the global deviation from the spatiotemporal effect as a whole. Since the spatial and temporal components of the whole adequately capture most of the structure, substantial spatiotemporal effects are not common. Therefore, we choose a hybrid model to calculate the distribution of spatiotemporal effects; ε_{it} is composed of two parts: the first part of the parameters only reflect the residual noise, while the second part captures the main effect of deviation from space and time. When heterogeneity is suspected, mixed models are usually used for Bayesian inference because they give a flexible prior structure that can be used for classification. Based on this, we construct the space-time effect as:

$$\varepsilon_{it} \sim \beta \text{Nomal}\left(0, \tau_1^2\right) + (1 - \beta)\,\text{Nomal}\left(0, \tau_2^2\right) \tag{5}$$

For formula 5, the Dirichlet $\beta \sim D(1, 1)$ distribution is assigned to the weight parameter $\beta = (\beta, 1 - \beta)$. To avoid the label switching problem [5], we define the normal semi-normal transcendental of the standard deviations τ_1 and τ_2:

$$\tau_1 \sim \text{Nomal}\,(0, 0.01)\ \ I_{(0,+\infty)}$$

$$\rho \sim \text{Nomal}\,(0, 100)\ \ I_{(0,+\infty)}$$

$$\tau_2 = \tau_1 + \rho \tag{6}$$

where I is the indicator functions. The formula in Formula 6 ensures that τ_2 always corresponds to the variance of the second component in all simulations, thus avoiding the label switching problem.

For intercept α, specify that it follows a normal distribution. The parameters are then assigned to the third layer in the third layer model. As Gelman [10] suggests, we specify a uniform prior value of 0 to 10 on the standard deviation scale of the variance parameters involved in second-order equations.

2.3 *Stability Judgment Based on Space-Time Effect*

As with the usual Bayesian mixture model, we define a potential assignment variable:

$$T_{it} = \text{Multinomial}\,(1, \beta) \tag{7}$$

The value 0 or 1 of T_{it} depends on whether we are analyzing the first component of the space-time effect Nomal$\left(0, \tau_1^2\right)$ or the second component of the space-time effect Nomal$\left(0, \tau_2^2\right)$. The second component of space-time interaction effect captures the main influence of really deviation from space and time, and the judgment of space-time stability depends on the second component of space-time effect. Therefore, we focus on the posterior probability p_{it} of the second component of the spatiotemporal effect:

$$p_{it} \equiv \Pr\,(T_{it} = 1|\text{Data}) \tag{8}$$

For the posterior probability p_{it}, we divide the stability of each region by two decision rules proposed by Abellan et al. [4]. We divide all regions into two categories: “stability regions” and “instability regions.” The rules of its division are as following:

Rule 1: In region i, if at least one p exists $p_{it} > P$, it is considered that this region belongs to “unstable region”; otherwise, it is “stable region.”

Rule 2: In region i, if the average of the three maxima of p_{it} is greater than P, the region is considered to be an “unstable region.”

For the threshold P, Abellan et al. [4] have shown that the critical value of 0.5 can balance the two types of errors (false-positive rate and false-negative rate) and guarantee good specificity.

3 Conclusion

3.1 Model Implementation and Convergence

The posterior distribution of Bayesian hierarchical model is usually difficult to calculate. In this paper, the posterior parameters of our model are implemented by WinBUGS software. The opening source software is based on a framework simulation of the Markoff’s Monte Carlo method, which combines Gibbs’s sampling with the metropolis Hastings step to estimate the posterior distribution of model parameters. In this paper, we simulate the model with two chains with different starting values, and after discarding the first 200,000 iterations of aging, we calculate the 200,000–250,000 posteriori results. The convergence was evaluated by the Gelman-Rubin convergence statistics, and Gelman and Rubin [9] concluded that for any of the model parameters representing convergence, the diagnostic statistic was less than 1.20. In this study, all model parameters were less than 1.05.

3.2 Model Performance Test

In order to test whether the Bayesian space-time hierarchical model can be overfitted and deviate from the actual results, we establish a Bayesian main space hierarchical model; the space effect calculated by main space model is compared with the space effect calculated by space-time model, whether the result of the spatiotemporal effect model will cause the loss of precision.

So, we establish a main space model with only space effects:

$$\log(Q_{it}) = \alpha + \zeta_i$$

$$\zeta_i \sim N\left(\mu_i, \sigma_\zeta^2\right)$$

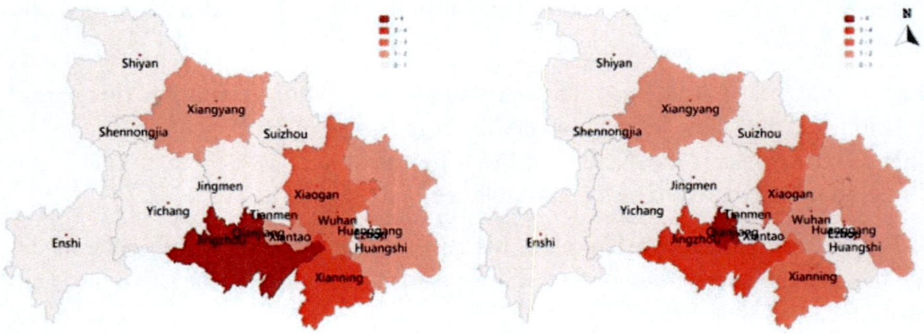

Fig. 1 The left is the spatial effect level of each region's main spatial model, and the right is the spatial effect level of the spatiotemporal model

$$\mu_i \sim \text{CAR}\left(W, \sigma_\mu^2\right) \tag{9}$$

By calculating the posterior probability $\exp(\zeta_i)$ of main space model and comparing the value $\exp(\zeta_i)$ of main space model (Eq. 2) with that of space-time model (Eq. 9), the results showed that the estimated value of the two models, $\exp(\zeta_i)$, was slightly different and the correlation coefficient was 0.9998. Figure 1 depicts the spatial effect levels in the main spatial model and the spatial effect levels in the spatiotemporal model for each region, respectively. From digital results and the visual results, it can be concluded that the accuracy of the spatial effect value could not be lost when the spatiotemporal model is used.

3.3 Results

The model was iterated over 250,000 times in WinBUGS software, and after abandoning the first 200,000 iterations, we analyzed the data generated after the last 50,000 iterations. Table 1 describes the standard deviation, variance, MC error, 2.5% confidence interval, and a posteriori value of 97.5% confidence interval for each parameter and hyperpriors in the Bayesian hierarchical spatiotemporal model.

For the posterior values of all parameters and their related parameters, it can be seen that the MC error of the parameters is much lower than the posterior standard deviation and the posterior values of the 2.5% confidence interval and the 97.5% confidence interval are close to the posterior standard deviation; this shows that our model has a good convergence. The posteriori values of accurate and stable, with credibility. So, the results can be further analyzed. The confidence interval of mixed parameter is relatively narrow, which shows that the mixed model of space-time effect has enough information to estimate.

For the space effect ζ_i in the space-time model, we calculate the posterior distribution $p(\exp(\zeta_i)|\text{Data})$ of the space effect. In the 17 cities in Hubei province,

Table 1 A posteriori summary of standard deviation of spatiotemporal models with 95% confidence intervals

				Credible interval		
Parameter	Mean	SD	MC error	2.50%	Median	97.50%
$\sigma(\zeta)$	0.8852	0.3789	0.0084	0.1912	0.8864	1.6520
$\sigma(\mu)$	1.4091	0.7352	0.0154	0.1529	1.3991	2.9101
$\sigma(\omega)$	0.0109	0.0053	1.62E-04	0.0028	0.0102	0.0232
$\sigma(\gamma)$	0.0816	0.0061	4.99E-05	0.0706	0.0813	0.0946
τ_1	0.0090	0.0058	2.63E-04	0.0025	0.0079	0.0247
κ	0.5693	0.0522	0.00079	0.4776	0.5656	0.6820
τ_2	0.5786	0.0516	6.91E-04	0.4885	0.5749	0.6898

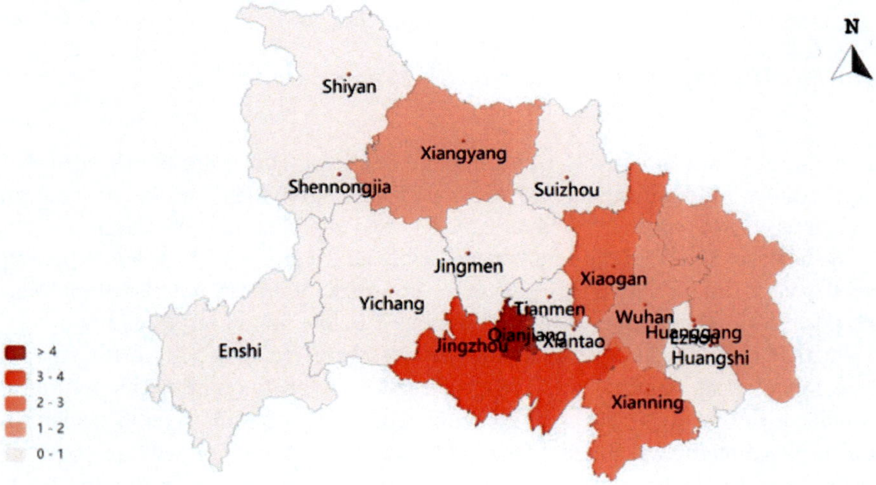

Fig. 2 The spatial effect of the spatiotemporal model

about 59% (10 cities) have spatial effect less than 1. About 41% of the cities (seven cities) had a spatial effect greater than 1. The spatial effect greater than 1 indicates that the urban travel volume has high activity, which also means that the social distance of these cities is relatively small. Previous studies have shown that the social distance is negatively correlated with the spread of COVID-19 [10]. Therefore, the higher the spatial effect of new coronavirus in the city, the faster the spread of new coronavirus, which is not conducive to epidemic prevention and control. City with high spatial effects should adopt a more stringent social distance intervention policy when facing COVID-19 (Fig. 2).

In fact, Hubei province did make strong social distance interventions after the discovery of the COVID-19. The map of a posteriori probability $p(\exp(\omega_t)|\text{Data})$ of time effect in the spatiotemporal model is shown in Fig. 3. After January 20, 2020, the posterior distribution of time effect decreased rapidly to about 0.5 and lasted for nearly one month after January 20, 2020, which reflects that the overall

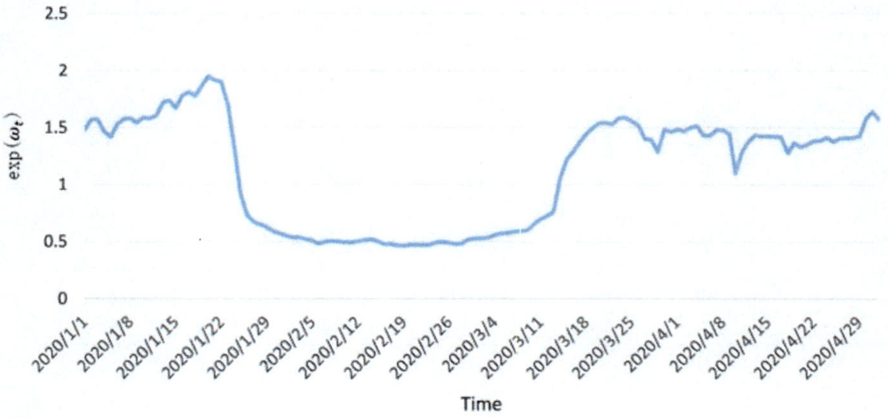

Fig. 3 Time effect trend chart of spatiotemporal model

passenger flow was effectively controlled after cities in Hubei province issued the city shutdown order on January 20. After mid-March, the lifting of the closures in most cities allowed traffic to gradually recover and approach January levels.

The posteriori value of the parameter τ_1 of the mixed model of the spatiotemporal effect is very small relative to τ_2, which is consistent with our initial assumption. The posteriori value of the weight coefficient β of the mixed model is 0.95; it is shown that most of the spatiotemporal effects ε are the first part of the mixed model; they are the normal small range errors of the spatiotemporal model; only a small part $(1 - \beta)$ of the spatiotemporal effects ε_{it} captures a large deviation from the spatiotemporal model. This is also in line with reality because most of the reality of the daily traffic flow is out of a stable state; only a small number of areas will be unstable state. We focus on the spatial-temporal effects of the second part of the classification of urban stability. The stability of our region is classified by the threshold $p = 0.5$. The results show that both rules found in Wuhan were the only "unstable region," and this kind of instability mainly displays in 2020.01.01–2020.01.20. That is, to say, the flow of travel was very volatile before the closure of Wuhan, coupled with Wuhan's own large flow of travel, making social distance and increased frequency of contact. We speculate that this may be one of the important reasons why the cumulative number of confirmed cases of the new coronavirus in Wuhan is much higher than in other cities.

4 Discussion

Based on the Bayesian hierarchical spatiotemporal model and the opening source data of travel intensity from Baidu, this paper analyzes the change and evolution of the travel intensity in Hubei province during the outbreak of COVID-19. According to the posterior distribution of the spatio-temporal model, we can see that the spatial effects are different in all fields, which indicates the spatial forces under the combined influence of the number of urban population and the intensity of travel in surrounding cities. Those cities with high spatial effect should implement more strict social distance intervention policy when facing the COVID-19. The time effect trend of Hubei province as a whole reflected that during the epidemic period, the strong measures of city closure did reduce the travel volume.

The spatiotemporal effect by means of the spatial effect and the time effect depicts the deviation between the daily spatial effect and the time effect of each city and the travel intensity. When there is a big deviation in the city, it means that there is instability in the city. We have detected that Wuhan has been in a state of fluctuating population since January 1, 2020 and this state of instability continued until the closure of the city, which accelerated the spread of the COVID-19. Instability can lead to sudden changes in the flow of people in public areas of cities which is clearly detrimental from the point of view of epidemic prevention and control. Changes of travel volume make it harder for close contacts to track. at the same time, the quantitative traffic control is more difficult to achieve than the unstable region and more difficult to determine than the implementation of sealing and unsealing strategy.

For the threshold P, Abellan et al. [3] have shown that the critical value of 0.5 can balance the two types of errors (false-positive rate and false-negative rate) and guarantee good specificity. We haven't looked at the threshold much further. In addition, our study is limited to exploring the local patterns of population fluctuations, without in-depth discussion of closely related indicators. Therefore, spatial or time-varying coefficient models can be further considered, to allow an explanation of the variable's spatially or temporally different effects on population fluctuations.

The construction of spatiotemporal model provides a perspective for epidemic prevention and control. When facing infectious diseases similar to COVID-19, cities with high spatial effect and cities in an unstable state should be paid more attention. This paper simply considers the relationship between spatiotemporal model and epidemic situation. The impact of spatiotemporal hierarchical model on epidemic transmission can be further discussed.

Acknowledgments This research is funded by the Natural Science Foundation of Inner Mongolia Autonomous Region under Grant 2020MS06021, Talent Development Foundation of Inner Mongolia, National Natural Science Foundation of China (71961022), and Graduate Scientific Research Innovation Project of Inner Mongolia University of Finance and Economics (NCYX2020-2019).

References

1. Kubíček, P.: Population distribution modelling at fine spatio-temporal scale based on mobile phone data. Int. J. Digital Earth. **12**(11), 1319–1340 (2019)
2. Shan, J., Ferreira, J., Gonzalez, M.C.: Activity-based human mobility patterns inferred from mobile phone data: a case study of Singapore. IEEE Tran. Big Data. **3**(2), 208–219 (2017)
3. Cao, J., Wei, T.U., Qingquan, L.I.: Spatio-temporal analysis of aggregated human activities based on massive mobile phone tracking data. J. Geo-Inf. Sci. **19**(04), 467–474 (2017)
4. Abellan, J.J., Richardson, S., Best, N.: Use of space–time models to investigate the stability of patterns of disease. Environ. Health Perspect. **116**(8), 1111–1119 (2008)
5. Zhensheng, W.: A Bayesian spatio-temporal model to analyzing the stability of patterns of population distribution in an urban space using mobile phone data. Int. J. Geogr. Inf. Sci. **35**(1), 116–134 (2021)
6. Banerjee, S.: Hierarchical Modeling and Analysis for Spatial Data, 2nd edn. CRC Press/Taylor & Francis Group, Boca Raton (2015)
7. Andrew, B., Lawson, S.: Bayesian Disease Mapping: Hierarchical Modeling in Spatial Epidemiology, 2nd edn. CRC Press, Boca Raton (2013)
8. Wei, T.: Coupling mobile phone and social media data: a new approach to understanding urban functions and diurnal patterns. Int. J. Geogr. Inf. Sci. **31**(12), 2331–2358 (2017)
9. Gelman, A.: Prior distributions for variance parameters in hierarchical models (comment on article by Browne and Draper). Bayesian Anal. **1**(3), 515–534 (2006)
10. López, L., Rodó, X.: A modified SEIR model to predict the COVID-19 outbreak in Spain and Italy: simulating control scenarios and multi-scale epidemics. Results Phys. **21**, 103746 (2020)

A Review of Additive Manufacturing (3D Printing) in Aerospace: Technology, Materials, Applications, and Challenges

XinXin Fu, YuXuan Lin, Xue-Jie Yue, XunMa, Boyoung Hur, and Xue-Zheng Yue

1 Introduction

Additive manufacturing (AM) is also called 3D printing [1–4]. It is a process of superimposing layers of "printed materials" with the control of a computer and turning the blueprint on the computer into a physical object [5]. This product is accomplished by a 3D printer under the operation of CAD software [6, 7]. The craft is characterized by fast, simple, and material saving. Above all, it can also quickly fabricate intricate and sophisticated components. Therefore, it is also known as rapid prototyping [8]. This technology has been employed for rapid prototyping over a long period of time [9–16]. Nowadays, as the sphere of application expands to include tissue engineering [17–19], chemistry reactors [20, 21], and electronics [22–24], this technology is no longer dedicated to prototyping [25].

Digital manufacturing technologies have attracted a lot of attention in recent years. Nowadays, a growing number of cardinal industrial countries in the world are promoting 3D printing technology as the basis of future manufacturing industry. Additive manufacturing technology began in the 1980s [4, 26, 27]. It has a history of 40 years at present. The conception of manufacturing objects by 3D

X. Fu · XunMa · X.-Z. Yue (✉)
University of Shanghai for Science and Technology, Shanghai, China
e-mail: Usst-yzyz@usst.edu.cn

Y. Lin
Anyang Normal University, AnYang City, South Korea

X.-J. Yue
XinXiang University, XinXiang City, China

B. Hur
Geyongsang national University, Jinju, South Korea

D. Tang et al. (eds.), *Mobile Wireless Middleware, Operating Systems and Applications*, EAI/Springer Innovations in Communication and Computing,
https://doi.org/10.1007/978-3-030-98671-1_6

printing technology originated from the late 1960s. Researchers at the Barthel Memorial Institute in the United States utilized the interaction of laser beams and photopolymers to obtain the desired objects in a resin cylinder. It is considered as the craft that manufactures a battery of objects utilizing the 3D model data [28]. This craft has been expanded by Charles Hull who discovered it in a course called stereolithography [27, 29]. Since the first 3D printing system was reported to the public, it has been making prominent progress until now [30–32]. Plenty of developments including fused deposition modeling (FDM), inkjet printing, and powder bed fusion are exploited, all of which are used up to now [33]. However, different from traditional manufacturing technologies that manufacture products by dislodging materials from larger raw materials or sheet metals, 3D printing manufactures the ultimate shape by joining materials, so that it can realize the maximum utilization of materials and desirable accuracy. In a great many niche industries at present, additive manufacturing is employed in combination with traditional manufacturing, such as subtractive manufacturing. It incrementally prompted components companies to appear [34]. Owing to the particular features of additive manufacturing, such as efficient and customizable fabrication, this technique is extensively employed in the domains of medical, electronics, aerospace, automobile, and so on [35, 36].

Above all, aerospace applications are proved to be one of the cardinal applications of additive manufacturing technologies. Among diverse 3D printing technologies, selective laser melting (SLM), selective laser sintering (SLS), electron beam melting (EBM), fused deposition modeling (FDM), etc. are the common processes in the aerospace industry. The aerospace industry is one of the cardinal application domains of 3D printing technology in prototype, testing, and product manufacturing.

In the aerospace industry, additive manufacturing is incrementally utilized in the manufacture of disparate individual aircraft components. One of the primary reasons is to redesign and manufacture components to meet the requirements of reducing mass and cost without affecting the mechanical properties of components. The significant feature of additive manufacturing technology based on fusion is rapid melting, which is able to produce extremely fine grains. Compared with traditional technology, it has superb advantages. In addition, this technology can also control the microstructure characteristics by manipulating process parameters to adapt to the more complex working environment and fabricate the lightweight structure products. Therefore, additive manufacturing is not only a prototype design method but also a direct manufacturing craft to fabricate high-quality near net shape products.

Additive manufacturing technology has made a great contribution in the domain of aerospace, whether in energy and resource saving or in environmental protection. However, 3D printing technology is still a developing technology. Due to the lack of established standards and certification for components produced by additive manufacturing, most of the current use is limited to non-mission-critical applications in the aerospace industry [37]. To address these issues, manufacturers and aviation regulators are incrementally working to develop new standards to satisfy the current capabilities of 3D printing. At the same time, additive manufacturing is also facing unprecedented technical challenges, which require a lot of related research.

This paper principally reviews the main processes, materials, applications, and challenges of additive manufacturing in the aerospace industry. The purpose of this review is to provide the latest advances in additive manufacturing in aerospace. The paper will be beneficial to university professors, research scholars, industrial experts, and entrepreneurs.

2 Main Processes in Aerospace

The aim of additive manufacturing processes is to establish and integrate layers in diverse ways. On account of the mechanism that each layer is formed, these processes can be separated into seven classifications involving adhesive spray, material extrusion, directional energy deposition, material spray, powder bed fusion, thin plate lamination, and vat photopolymerization [38–41]. Furthermore, 3D printing can be grouped into two categories, i.e., (i) in view of the physical state of the raw material, that is, based on solid, liquid, or powder, and (ii) according to the way in which substances are melted at the molecular level, that is, heat, ultraviolet, laser, or electron beam [1, 2].

Among these additive manufacturing processes, there are several main technologies in aerospace which are concisely concluded as follows.

2.1 Fused Deposition Modeling (FDM)

FDM is a sort of material extrusion process to manufacture thermoplastic components through heated extrusion and layer-by-layer deposition of materials [42]. The overall manufacturing process of FDM primarily includes material preparation, processes of 3D printing, and post-operation. The methodology of FDM printing process consists of the following steps. At first, the 3D model to be printed is drawn by CAD software. Subsequently, the model needs to be exported in STL format. Afterwards, the STL file is imported into the slicing software, and the model is sliced according to the set parameters. And then, the 3D printer starts to work and prints the model after slicing layer by layer. Eventually, the desired 3D solid model is to be got. Figure 1 is the manufacturing process of fused deposition modelling (FDM).

This technique is commonly used to fabricate polymer matrix materials. FDM fabricates polymeric components from polymer thermoplastic by extruding from the nozzle and then melting and depositing it by adjusting the diversified processing parameters. For the sake of thoroughly developing the FDM process, each link should be gradually improved to achieve the goal of perfection.

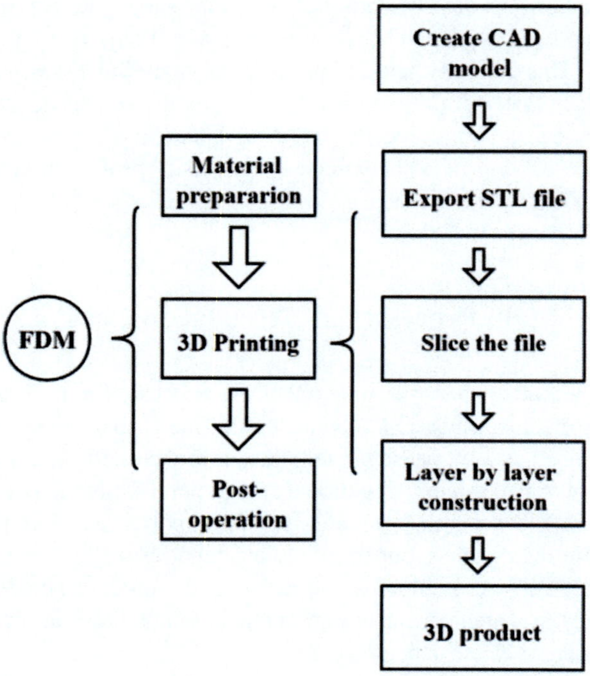

Fig. 1 Manufacturing process of fused deposition modeling (FDM)

Based on the fused deposition modeling, Matsuzaki et al. [43] developed a novel method by continuous fiber-reinforced thermoplastics. This way has the advantage to save materials and printing time without making the mold, which will become the standard process of manufacturing composites in the future. In this process, polylactic acid was used as the matrix, while carbon fibers, or twisted yarns of natural jute fibers, were used as the reinforcements. With respect to the conventional additive manufacturing polymer matrix composites, continuous fibers are supplied to the raw materials, most importantly, impregnating the fibers with filaments within the heated nozzle before printing, which has vastly improved the tensile strength of composites. Zaman et al. [44] put forwards the parametric optimization of the FDM process taking advantage of the Taguchi design of experiments. These experiments were carried out under the drilling grid from the aerospace industry to research the impact of FDM parameters including thickness, shell, filling method, and infill percentage on the compressive strength of objects. The result was that infill percentage was the leading factor for the objects. In addition, the optimal combination of process parameters to maximizing the compressive strength was found to like the selection of levels for the approved parameters of aerospace industry. Gebisa et al. [42] investigated the influence of FDM process parameters on the flexural properties of expected materials. The full factorial design experiment including contour width which was a novel parameter was implemented using the UL TEM 9085 material. According to the research, the raster angle and raster width were the most dominant factors, and the secondary factors were the contour number and contour width.

2.2 Selective Laser Melting (SLM)

Selective laser melting (SLM) is deeded to one of the most prospective additive manufacturing technologies. SLM is a process of powder bed fusion [45]. There is no need for the 3D product manufactured by SLM to do post-operation on the whole. Therefore, the overall manufacturing process of SLM basically includes material preparation and processes of 3D printing. The methodology of SLM printing process involves the following steps. Primarily, the 3D model is drawn by CAD software such as SolidWorks, UG, AutoCAD, and so on. Generally, the 3D model is exported in STL format and then is to be sliced. Subsequently, a high-energy laser beam is to melt pre-spread powder. After it solidifies, repeat the operation of spreading and melting powder until the desired component is printed completely. The framework diagram of the methodology of SLM is shown in Fig. 2.

Sing et al. [46] researched the influence of different process parameters on the dimensional accuracy and compressive behavior of cellular lattice structures manufactured by SLM. It was reported that the laser power and speed of laser

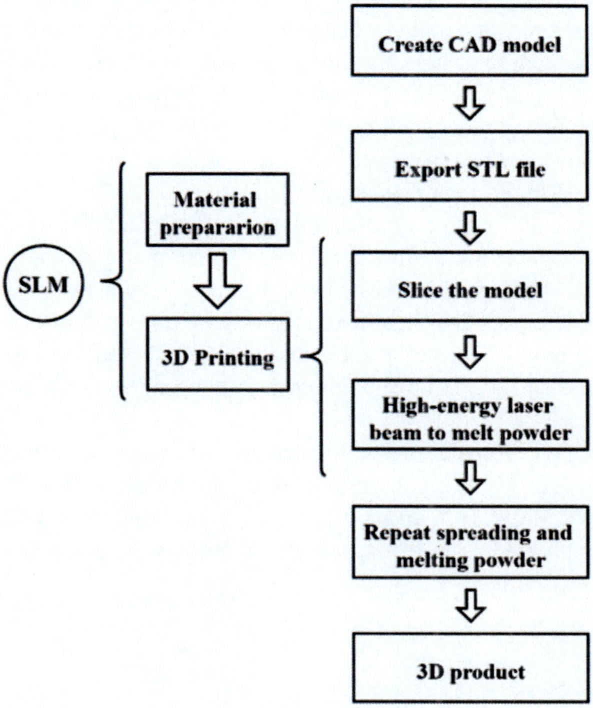

Fig. 2 Manufacturing process of selective laser melting (SLM)

scan had little influence on the elastic constant, while they affected significantly the dimensional accuracy. Nevertheless, the type of unit cell and structure diameter distinctly impacted the elastic constant, but not dimensional accuracy. A large number of investigations indicate that aluminum and titanium can conform to the requirements of lightweight in the aerospace field. For example, Spierings et al. [47] have analyzed the hardness response of different heat treatment temperatures for Sc- and Zr-modified Al-Mg alloy (Scalmalloy®) manufactured by SLM. The results indicated that the heat treatment faintly impacted the mechanical properties with Rm-values exceeding 500 MPa. Shipley et al. [48] focused on how to eliminate the effects including defects and residual stresses of Ti-6Al-4V fabricated by SLM which were a result of post-processes. For the sake of overcoming the harsh environment in space, Metal Matrix Nanocomposites (MMNCs) are widely concerned as they have outstanding properties with high specific stiffness and near-zero CTE to satisfy the demand for lightweight in the aerospace industry. Whereas the manufacturing process and economic cost have limited the extensive application of MMNCs to a great extent. SLM has exhibited prospective performance to work out the challenges of MMNCs [49].

3 Materials

Aviation structural materials primarily refer to the materials used in aircraft, airframe, and engine. The engine material is the most important structural material in aviation materials. Aluminum alloy, titanium alloy, high-strength steel, and so on account for a large proportion of aviation structural materials, but they also face more and more challenges from polymer composites, metal matrix composites, and non-metallic materials.

Aircraft materials are divided into airframe materials (including structural materials and nonstructural materials), engine materials, and coatings, the most important of which are airframe structural materials and engine materials. Nonstructural materials include transparent materials, cabin facilities and decoration materials, accessories and pipe materials for hydraulic and air conditioning systems, etc. The amount of nonstructural materials is small, and there are many varieties, including glass, plastics, rubber, aluminum alloy, magnesium alloy, copper alloy, and stainless steel.

The comprehensive requirements of aerospace structural components concentrate on lightweight, prominent mechanical strength, high impact damage resistance, etc. [50]. The primary materials that accord with the additive manufacturing requirements in aerospace are as follows: metals and alloys, ceramics, polymers, and composites.

3.1 Metals and Alloys

Additive manufacturing requires varied materials. With the diversification of demand and the development of technology, a variety of new materials has emerged endlessly. A large number of studies are in progress on employing diverse forms of alloys for additive manufacturing. The minimum thickness of the printed layer including polymer, ceramic composite, and aluminum alloy is 20–100 μm, decided by the additive manufacturing process and the physical condition of the material [51]. Familiar metal materials are employed for manufacturing aerospace components involving tool steel and stainless steel, titanium, nickel, aluminum, and alloys of these materials. Also, gold, platinum, and silver are employed for specific applications in the aerospace industry [5]. Nevertheless, Ti-based and Ni-based alloys are more significant in the aerospace industry [52, 53]. Especially, nickel-based superalloys have been incrementally applied, as a result of their extraordinary performances at high temperatures. They are bitterly appropriative for aerospace components working in radical environments [54, 55]. In the aerospace domain, Ni-based alloys are highly favored, because of their tensile properties, damage tolerance, and anticorrosion or inoxidizability. Though these alloys lead to the tendency of high cracking, the mechanical properties of them can be improved by HIP craft [53]. Besides, an unprecedented bimetallic structure utilizing two disparate aerospace alloys can tremendously amelioration its thermophysical properties compared with nickel alloy. This type of structure has two characteristics consisting of shape memory effect and hyperelasticity making them broadly applied. Since separate process parameters immensely affect the thermal and mechanical properties of NiTi alloys, Mehrpouya et al. [56] proposed a machine learning algorithm of artificial neural network to predict the optimal process parameters of additive manufacturing. A great many researches have demonstrated that this algorithm has a favorable prediction effect, which can be widely used. It can be seen that multi-material structure can work out various performance defects of single material.

Moreover, the Ti-6Al-4V alloy has extensively attracted the attention of the aerospace industry, as a result of its incomparable properties involving high intensity and fracture toughness, low density, and thermal expansion coefficient [57–59]. Besides, as a lightweight material, its high corrosion resistance is charismatic to aerospace structures [60]. While Ti-6Al-4V fabricated by additive manufacturing represents outstanding merits, there are a large number of challenges to be settled in order to substantially be utilized in the aerospace industry. Especially, the fatigue properties of Ti-6Al-4V are the primary matter. Kahlin et al. [61] investigated the fatigue properties of Ti-6Al-4V manufactured by SLM and EBM. The results show that surface roughness is the principal factor. Therefore, for the sake of ameliorating the fatigue strength of SLM and EBM materials, surface post-processes such as hot isostatic pressing need to be utilized to reduce the severity of microcracks.

3.2 Polymers and Composites

It is well known that polymers and composites are the most common materials used in additive manufacturing. The first developed 3D printing technologies are taking advantage of manufacturing pure polymers. In the midst of the technologies, FDM is the most frequently utilized process due to low cost and wastage. However, pure polymers typically display low mechanical properties. Therefore, composite materials have increasingly attracted attention on account of the excellent mechanical properties.

Ning et al. [62] presented the carbon fiber-reinforced thermoplastic composites fabricated by FDM. Compared with pure polymers, appending carbon fiber into plastic could enhance the tensile strength and flexural properties. Continuous fiber-reinforced materials have better mechanical properties generally used in all sorts of applications relative to short fiber-reinforced materials. An innovative additive manufacturing process fused filament fabrication (FFF) is proposed, which can be the desired technology to carry out prototyping and customization [63]. In addition, it is found that the properties of materials are also related to the interfacial characteristics through the mass investigations on continuous fiber-reinforced thermoplastic polymer composites fabricated by additive manufacturing technology. The research on interface modification principally concentrates on ultrasonic treatment. Qiao et al. [64] fabricated continuous carbon fiber/polyacrylic acid composites using FDM, which innovatively proposed an ultrasonic penetration method to improve the interface properties. It is known that ultrasonic treatment only physically modifies the interface corresponding to fiber and resin. Moreover, the tensile strength and bending strength of the composites are enhanced. On account of the increasing requirement in the aerospace industry, the composites of carbon fiber-reinforced materials combined with polyether ether ketone have intensively attracted attention. Polyether ether ketone is a kind of semi-crystalline thermoplastic material with excellent chemical inertia and high-temperature resistance. Luo et al. [65] investigated a plasma-laser co-treatment to optimize the interface between carbon fiber and polyether ether ketone matrix in composites manufactured by extrusion process. It is found that the laser enhances the interface bonding at the macro level, while the plasma improves the interface bonding at the microlevel. Furthermore, the effect of the microcosmic interface of plasma on performance is more distinct than laser.

The exploration of space has brought about a clearer understanding of the universe. Manufacturing strong and ultralight structural components that adapt to the extreme environments of space is needed to accelerate the pace of space exploration. Since the structure with a thermal expansion coefficient close to zero can ensure dimensional stability to a certain extent, numerous researches on carbon fiber-reinforced polymer are developed. Anguita et al. [66] invented a physical surface barrier which is a mechanically coupled enhanced carbon fiber-reinforced polymer so as to address this challenge that the entrance and release of water can bring about dimensional instability. This barrier immensely decreases the diffusion rate of water, more importantly, along with reducing surface contamination.

3.3 Ceramics

By comparison with metals and polymers, ceramics have extensively superb properties, involving high melting point and mechanical intensity and remarkable thermal stability [67]. When taking into account ceramic selections, alumina (Al_2O_3) and zirconia (ZrO_2) are profitable in aerospace applications. Owing to the superb thermal protection and mechanical properties of ceramics, it is of benefit to the aerospace industry [68].

The manned spaceflight program is an extremely crucial project in the exploration of space. To further implement the space program, engineers intend to establish a human outer space station. At this time, the combination of additive manufacturing technology is extraordinarily recommendable. Goulas et al. [69] put forward a powder bed fusion (PBF) process with a thermal energy source to fuse the particulates of ceramics multicomponent materials together to simulate the regolith of Lunar and Martian. The results exhibit that the regolith by powder bed fusion is more compatible with the Lunar rather than Martian. With regard to the Martian, Karl et al. [70] demonstrated the slip casting process based on water from Mars, which is an in situ resource utilization to establish a processing route for fabricating ceramics. They successfully manufacture the ceramics with steady mechanical properties, which may be a starting point of Mars colonization in the future.

Ceramics and ceramic-reinforced metal matrix composites are broadly applied in harsh working environments owing to their unprecedented chemical inertia and high-temperature resistance. Specifically, engineering ceramics are attractive for additive manufacturing aerospace applications thanks to their unrivaled high-temperature properties that can maintain excellent mechanical properties even in high-temperature environment. Moreover, there are also unique chemical and mechanical properties. Among them, zirconium dioxide has extremely high toughness, heat insulation, and ionic conductivity. However, the process of the materials is confronted with plenty of straits. An innovative processing route for fabricating additive manufacturing ceramics components is introduced [71], which has combined the superiorities of syringe extrusion and UV curing. There is the potential to manufacture multi-materials, tidy, and complex structural ceramic components.

Every production process of additive manufacturing corresponds to different aspects of different application fields, which primarily depends on the descriptions and characteristics of materials, processing methods of technologies, and performance requirements of diverse application domains.

4 Aerospace Applications

The aerospace industry is one of the significant application domains of additive manufacturing technology in designing a prototype, testing properties, and manufacturing finished components. Furthermore, the aerospace industry has fit into the 3D printing process from conceptual design to the employment and maintenance of components. The fields of applications consist of rapid prototyping of assemblies in the design gradation, subsequently manufacturing matrices or implements for duplicate work, straightway making aerospace components with complicated shapes, and restoring spoiled parts.

Stronger, lighter, and more durable assemblies are required in the aerospace industry. The comprehensive requirements of aerospace components principally focus on lightweight, superior mechanical properties, etc. Nowadays, additive manufacturing technologies are creating numerous new probabilities for coping with these challenges.

4.1 Unmanned Aerial Vehicle (UAVs)

In the aviation industry, unmanned aerial vehicles (UAVs) are increasingly utilized by numerous countries and industries. Their potential is constantly explored by scientists and entrepreneurs. Furthermore, lightweight UAVs are an extremely promising application field.

The build and print orientations are the key printing parameters for manufacturing unmanned aerial vehicle components by the fused deposition modeling. To analyze the ability of FDM-ed components for unmanned aerial vehicles, Ravindrababu et al. [72] furnished with a notion that evaluated the effects of build and print orientations on the FDM-ed unmanned aerial vehicle components by simulation. The mechanical properties of FDM components were evaluated and examined by comparing different build (edge-up (EU), face-up (FU), and straight-up (SU)) and print orientations (0–90°). The results showed that the stiffness and tensile strength of EU samples are the highest. In addition, the build orientation has a greater influence on the elastic deformation of FDM-ed components than print orientation.

Based on the small UAV framework, Azarov et al. [73] provided a novel method that the three-dimensional printing of continuous fiber-reinforced composite (CFRC), which took place of the fused deposition modeling process. Moreover, the frame has manufactured made of continuous carbon fibers and two kinds of matrix materials including thermoset and thermoplastic. CFRC has excellent merits that not only can rapidly prototype but also can obtain materials with less density and high mechanical properties such as high stiffness and strength. Therefore, the additive manufacturing CFRC can be deemed as an optimal substitute for small UAV components.

4.2 *Satellites and Rockets*

Aerospace applications may turn out to be the most considerable application of additive manufacturing technologies. Specifically, satellites and carrier rockets are confronted with radical temperatures ranging from liquid refrigerant to rocket engine combustion and evacuation. In the same way, the carrier rockets also have to endure high structure, vibration, and acoustic loads. High pressure and high speed exist in the combustion chamber, turbopump subassembly, and jet-propelled turbine. These technical challenges have hindered the exploration of the Unknown Universe. Consequently, a great many researchers are constantly exploring the breakthrough of rocket engines. Research results show that nickel-based superalloys, especially Inconel, have excellent creep properties, oxidation resistance, and heat corrosion resistance [54], so they can be widely used in turbine blades, combustion chambers, and other domains. For example, SpaceX adopts high-performance alloy Inconel to manufacture rocket engine parts.

SpaceX wants to advance 3D printing in the twenty-first century by manufacturing high-performance engine parts. They successfully tested the SuperDraco rocket thruster shown in Fig. 3, a 3D printing engine that powers the launch escape system of a spacecraft. The company's current version of Drago for reentry into the atmosphere uses Drago's modified engine for reentry. Its engine room is fabricated by direct metal laser sintering (DMLS). In addition, it is made of Inconel, a variety of high-performance superalloy, which is able to provide high strength and improve reliability [74]. SuperDraco is utilized to maneuver in orbit and reentry process. It will be used in the manned version of the Dragon spacecraft, as a part of the spacecraft launch escape system, and it will also be used to achieve land propulsion landing. Each SuperDraco has the capacity that generates 16,000 pounds of thrust. The eight SuperDraco engines installed on the sidewalls of the "Dragon" spacecraft will generate up to 120,000 pounds of axial thrust to transport astronauts to safety in the event of an emergency during launch.

4.3 *Aero-engine*

With regard to aero-engine, the ascending of working temperature will firsthand impact fuel efficiency. Additive manufacturing has the capacity to machine high-temperature materials, such as nickel alloys and intermetallic materials which are arduous to cast and process. These materials can also be employed at higher temperatures. The process can neatly manufacture intricate subassemblies with diverse shapes, ingredients, structures, and performances on the basis of the demands of designers, superseding the orthodox craft of processing parts [75].

As part of NASA's development of Mars exploration technology, NASA engineers have printed the first full-scale copper rocket engine component by additive manufacturing. The gradient lattice structure is shown in Fig. 4. It is designed

Fig. 3 The SuperDraco rocket thruster. (Reproduced from [74])

Fig. 4 3D printed rocket engine part. (Courtesy NASA)

by Amaero company. The lattice structure is extensively applied in the military, aviation, and astronautic industries. The lightweight and high-strength lattice structure would suit application in the aerospace industry.

This is a milestone in additive manufacturing aerospace. Additive manufacturing can reduce the time and cost of making rocket parts, such as copper liners found in rocket combustion chambers, where ultracold propellants are mixed and heated to the extreme temperatures required to launch the rocket into space. In the combustion chamber, the combustion temperature of the propellant exceeds 50,008 Fahrenheit degree. To prevent melting, hydrogen at temperatures below 100 degrees absolute zero circulates through the cooling inlet visible at the top edge of the combustor. In order to circulate the gas, more than 200 complex channels are built between the inner and outer walls of the combustor. The part is made of GRCo-84, a copper alloy invented by material scientists at the Glenn Research Center in Cleveland, Ohio, which helps verify 3D printing process parameters and ensure manufacturing quality [76]. Copper is an ideal material for rocket engine components on account of its excellent thermal conductivity. Simultaneously, it also brings rigorous challenges, since it is formidable to melt copper powder continuously by laser. Fortunately, Marshall's materials and processing lab resoundingly solved the problem by using a selective laser melting machine, fusing 8255 layers of copper powder in 10 days and 18 hours to make a combustion chamber. As a result, they created a sort of copper additive manufacturing process.

SmarTech has summarized four crucial ways that extract value from 3D printing in the aerospace domain, which are (1) abatement of production cost, (2) abatement of component weight, (3) abatement of lead time, and (4) abatement of the passive environmental impacts of production [77]. They are pivotal aspects of additive manufacturing in the aerospace industry. Hence, the future development of aerospace industry will also focus on these aspects as the premise for researching and manufacturing.

In the aerospace industry, numerous components are obliged to satisfy extreme working requirements. Correspondingly, comprehensive functions (i.e., structure, heat emission, and air current) are required in accordance with complex geometric structures. Complex structures demand more individual components, which must be fixed by nuts, bolts, and brazing. These junctions can abate the reliability of the components. Complicacy not only brings challenges to manufacturing technology but also makes traditional manufacturing costs rise sharply. Nevertheless, additive manufacturing components are able to be fabricated on demand, which can raise the degree of freedom of design, thus greatly reducing the growth of manufacturing costs. At the same time, the reliability can be improved by reducing the number of components and connection points [78]. For example, GE fuel nozzles have realized that parts can be simplified by combining multiple components.

GE holds a leading position in the application field of additive manufacturing of aircraft propulsion systems, integrating various technologies into new product development. More importantly, the company has resoundingly manufactured LEAP engine fuel nozzles shown in Fig. 5 by employing laser 3D printing fusion craft [79]. LEAP engine fuel nozzle has already passed the ground engine test and obtained the attestation of using civil aircraft. This is a landmark application. It is designed to power Boeing 737 MAX and Airbus A320neo aircraft [95]. Advantages of 3D printing fuel nozzles include a fivefold increase in durability, a

Fig. 5 Laser sintered leap engine fuel nozzle. (Courtesy GE Aviation)

25% reduction in body weight, and accelerated integration of components. The 3D printing assembly possesses the merits including abating cost and weight, without joints which have improved property.

4.4 Aerospace Flight

Apart from the problem of complexity, the development of the industrial structure chain also needs to consider the lightweight and low cost of products. This is also a key aspect of the additive manufacturing aerospace industry [77]. It can be solved by reducing the weight of the components. Therefore, some companies have made corresponding researches and manufacturing.

Lockheed Martin fabricates titanium alloy scaffolds that can be carried on the solar-powered Juno spacecraft by employing EBM [79]. Lockheed Martin is also looking to the other spacecraft projects by 3D printed. To employ additive manufacturing, Lockheed Martin makes a fresh start to refashioned an antenna reflector, which largely lessens the weight from 395 to 40 kg [80]. Boeing fabricates 3D printed plastic inner components. Those components composed of nylon are primarily aimed to manufacture archetypes and specimens. The company also fabricates molds that are aimed to produce composite components.

For the non-load-bearing parts of the aircraft, reducing weight is a quite promising approach for the aviation industry. The additive manufacturing technology can manufacture the lightweight hollow structure, which is arduous or even impossible to realize by traditional technology. For example, Sogeclair, a French airline supplier, has successfully produced lighter aircraft doors with additive manufacturing

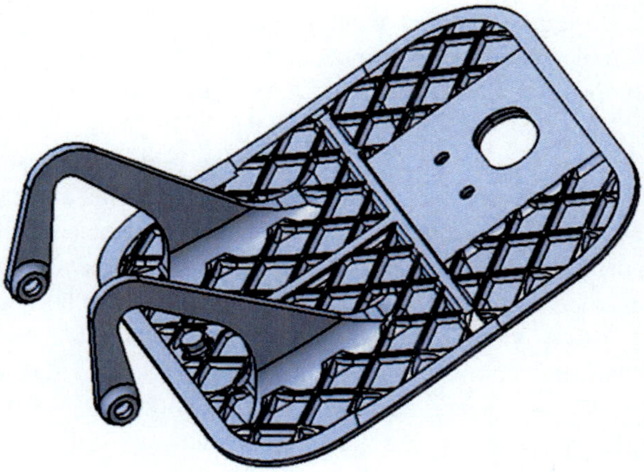

Fig. 6 Bracket of Ti-6Al-4V model

technology. Compared with the original door, the weight of the new door is reduced by 30%, and the strength remains unchanged. The process takes advantage of bionic network to optimize the model. The material adopted is PAAM, and the mold is printed by the large-scale 3D printer VX1000 of voxeljet in Germany. Finally, the molten aluminum is poured into the mold for casting. This craft has further promoted the development of the aviation industry.

A large number of research results show that titanium and its alloys have excellent comprehensive properties of high strength, high fracture toughness, low density, and high corrosion resistance [60]. In particular, Ti-6Al-4V alloy has high strength, high modulus, low expansion coefficient, and high corrosion resistance than aluminum alloy [59, 81], which is a sort of awfully attractive lightweight spacecraft structural material. It can be extensively employed in the aviation industry. Although the structure of the spacecraft is primarily made of carbon-/polymer-based composite materials, titanium alloys are utilized for a number of brackets, fittings, and support tubes. Currently, for the sake of fabricating complex-shaped brackets and accessories, the solid billet of titanium alloy is processed to the final configuration.

In 2014, Airbus manufactures the A350 bracket of Ti-6Al-4V shown in Fig. 6, which is the first metal 3D printing component employed in commercial aircraft [82]. The commercial airplane manufacturer Airbus has gradually attached the importance to laser melting of metal powders in airplane fabrication. The additive manufacturing Titanium Component in Airbus A350 XWB is made of titanium powder materials, leading to an over 30% reduction in weight. Its bracket is fabricated by employing laser focusing technology [75].

Researches show that shortening the lead time of components is the biggest source of value. Therefore, shortening the lead time of new parts and replacement parts may be the largest source of value for 3D printing in the aerospace domain in the next decade.

Currently, additive manufacturing has been diffusely employed in the manufacture of aerospace components, for instance, engines. The average lifetime of commercial aircraft is about 20–30 years, over time, and regular maintenance is indispensable [83]. The components of engines are light to be destroyed, which results in the periodical renewal. In the propulsion system of an aircraft, there are commonly more than 30,000 solitary components that demand regular maintenance [84]. To conform to the requirement, proficient companies provide maintenance, repair, and overhaul (MRO) services. The turnover period of MRO suppliers is regarded as a momentous performance indicator because it can retain the aircraft operation by minimizing the time of maintenance and maximizing the airline profits. Additive manufacturing can dramatically curtail the design and delivery cycle, which is very beneficial to MRO suppliers [85]. Hence, 3D printing technology is an unexceptionable means to settle this matter [86]. Legacy aircrafts are generally no longer in production; as a result, components renewal is difficult to achieve. The US Air Force (USAF) has established a partnership with America Makes to furnish on-demand manufacture for legacy aircraft and lessen the time for maintenance components. Low-volume manufacture brings about the decrease of components inventory; consequently, the company turned to on-demand manufacture [34, 87]. Except for minimizing inventory, the implementation of a 3D printing system can also cut down the cost of waste disposal.

In terms of environmental impact, a number of measures are implemented by reducing the weight of components. Of course, there is the reuse of raw materials to reduce the consumption of materials, so as to achieve the objective of green technology. Researchers believe that the future development of the aerospace industry will be more environmentally friendly since this will not only protect the environment but also accelerate the development of additive manufacturing technology.

4.5 Machine Learning

Since the widespread application of additive manufacturing in the industry has been hindered, machine learning (ML) is gaining increasing attention owing to its unparalleled performance in data integration. Depending on separate additive manufacturing techniques, the powder bed fusion (PBF) and material extrusion (ME) are primary processes in aerospace industry.

PBF utilizes a laser or electron beam as an energy source to fabricate components layer by layer through selectively melting metal or plastic powders. On the basis of disparate applications, PEF prevailingly encompasses selective laser sintering (SLS), selective laser melting (SLM), electron beam melting (EBM), etc. Among them, SLM and SLS make use of a laser as the energy source, while EBM takes advantage of an electron beam. Under the ME category, FDM is a representative process. Otherwise, the machine learning technologies can be universally divided into several categories: supervised learning, unsupervised learning, semi-supervised learning, and reinforced learning [88]. The application of ML algorithm in the aerospace additive manufacturing field primarily includes the design of 3D printing, the monitoring of the process, and inspection and evaluation of quality. The design of 3D printing occupies a dominant position on the process workflow, which is to improve the 3D printing technology from the source such as material design and topology design. The monitoring of the process is absolutely essential to carry out the optimization of printing technology as the additive manufacturing process itself still suffers from disparate defects. Through in-process monitoring of the AM process, the closed-loop feedback control can be implemented, which immensely improves the reliability of 3D printing. Quality inspection and evaluation is the final process, which is to ensure the quality of the finished product and further to complete the application of printed parts to meet the standard.

In the in situ monitoring of additive manufacturing, Zhang et al. [89] put forward a machine learning method for real-time evaluation of weld penetration defects on the basis of arc audible sound sensing made of aluminum alloy. Based on the arc voltage signal and the spectrum of weld defects, the wavelet filtering-based principal component analysis and a classification model embedded within parameter optimization and cross-validations are proposed, which boost the test accuracy to 98.46%. To ameliorate the uncertainty of the quality of additive manufacturing products, Okaro et al. [90] come up with a semi-supervised machine learning algorithm to automatically detect product faults. By collecting data from photodiode sensors and extracting key features, the Gaussian mixture model is trained to identify defects in the printing process. Li et al. [91] introduced a machine learning modeling approach to predict surface roughness so as to further improve the surface integrity which limited the development of additive manufacturing. Real-time monitoring data is collected through multiple sensors. An ensemble learning algorithm combining six various algorithms including RF, AdaBoost, CART, SVR, RR, and RVFL is present to accurately predict the surface roughness of 3D printed samples. Furthermore, Mukherjee et al. [92] furnished a comprehensive digital twin machine learning algorithm. The method has the superiorities to diminish the volume of trial and error and lessen the time of product manufacturing.

5 Superiorities

In contrast to the traditional manufacturing process, additive manufacturing has a great many superiorities. In terms of the utilization of resources and materials, additive manufacturing processes make efficient use of raw materials and also simplify the processing of surplus materials for reuse [93]. At the design level, additive manufacturing consists of freedom and flexibility, most importantly, the ability to customize components on the basis of your requirements, which is unparalleled by traditional processes. In printed products, additive manufacturing pays more attention to implement the lightweight and high performance of products through designing the porous structure of the material.

6 Challenges

Although additive manufacturing has a number of unrivaled advantages, it still is confronted with inevitable challenges in various aspects. This section chiefly elaborates on challenges from additive manufacturing in the aerospace industry.

Additive manufacturing has gradually expanded from rapid prototyping to rapid mold, direct components manufacture, and maintenance. Additive manufacturing technology has increasingly displayed superb ability and potentiality in the domain of aerospace applications. With regard to the emerging industry of additive manufacturing in aerospace, there are still challenges to be addressed, whether in the design of materials including structural design and topological design or in the manufacturing process involving equipment transformation, process parameters, and energy utilization. The cardinal challenges are summarized below.

In terms of the processes, the challenges primarily come from the process itself and the application. For the UAVs fabricated by FDM, the major researches are still in the initial stage. Experiments show that the 3D printing process parameters have a remarkable influence on the mechanical properties of UAVs. The components fabricated by FDM have the traits consisted of anisotropy and void formation, which will have a severe impact on the mechanical properties of the parts [94]. The building orientation of FDM will affect the surface roughness of components. In addition, the support structure is also a limitation of FDM, which increases the consumption of time and materials [95]. Optimizing 3D printing process parameters is required to improve the mechanical properties to a certain extent, such as controlling layer thickness, build orientation, temperature, printing speed and other parameters, and running intelligent algorithm by training data to realize optimal design [96].

In terms of the materials, the properties and structure of materials have a mysterious space for transformation. In outer space, the semifinished material of 3D printing metal isn't allowed as powdery, due to its palatability. NASA's Langley Research Center points out that EBF3 can be employed to resolve this matter. The technology employs an electron beam gun to manufacture the 3D printing components [97].

Currently, additive manufacturing is still unable to completely replace traditional manufacturing. Particularly, there are obstacles in the field of mass production. For large-sized objects, additive manufacturing processes lack the ability to satisfy the demand for strength, which has a bad effect on the surface finish of the components [93]. The application of additive manufacturing in spacecraft principally in space is faced with a great many technical difficulties to breakthrough. Optimization design has not been perfectly matched with 3D printing technology, and it still demands to rely on traditional methods for design. Topology optimization combined with 3D printing technology not only meets the requirements of high performance but also realizes the lightweight design of structure, which is an excellent breakthrough. Moreover, it has been proved that thermoelastic topology optimization combined with SLM 3D printing technology can realize thermomechanical load [98].

In addition, the interface combination of various materials and the manufacturing of complex structures are in urgent need of technical breakthroughs. Though additive manufacturing supplies incomparable superiorities for the aerospace industry to manufacture difficult to process materials, within the industry, it is impossible to supply a standard database of material properties for mechanical properties manufactured by diverse 3D printing processes, such as the fatigue response under dynamic load. A large number of researches show that porosity, residual stress, and surface roughness are crucial elements for the fatigue properties [99], affecting the mechanical properties. Appropriate posttreatment and HIP can improve it. HIP, in particular, has been widely implemented in the field of aerospace [100].

7 Prospects

This section chiefly elaborates on new possibilities from additive manufacturing technologies in the aerospace industry.

In the domain of aerospace, there are a great many opportunities for additive manufacturing technology. Starting from the most fundamental raw materials, the application of high-strength lightweight materials in aerospace will bring about enormous economic effectiveness [40]. Aerospace companies are investing emphatically in additive manufacturing applications, especially General Electric. From the perspective of sustainable development, additive manufacturing has the superiorities of high utilization of materials and low material waste [34]. In the face of global climate change, governments and aviation organizations have set targets to lessen carbon dioxide emissions in the near years and have issued a sequence of regulations for realizing this goal. To reach these objectives, aerospace companies are incrementally endeavoring to make good use of 3D printing technology, to cut down fuel consumption and advance the property efficiency. Aerospace is one of the most suitable domains to adopt additive manufacturing technology as a part of enterprise value stream.

The effectiveness and potential of additive manufacturing are being incrementally discovered. A good many organizations, universities, and national consortia work together, in order to raise the most advanced level. The original intention of the development of aerospace industry is to continuously explore the unknown mysteries among the earth, the moon, Mars, and even the whole universe. Through the exploration of other planets, we can have more possibilities to be aware of the existence of unknown organisms, to constantly open up the relationship between human beings and the operation of the universe. We should strive to excavate the broader universe, recognize more profound mysteries, and lead mankind to a more advanced era. For the sake of coming true to these great visions, the assistance of additive manufacturing technology is indispensable.

Presently, the aerospace industry has developed in a number of applications, such as topological majorization, functional concordance, and straightforward manufacture of complex geometry and new material components. In the aerospace industry, 3D printing applications principally focus on the design, manufacture, and maintenance of aero-engine components. 3D printing is expected to have the following latent applications in the aerospace industry in the coming days:

UAVs Fabricated by FDM FDM is expected to become one of the core technologies in future industries for unmanned aerial vehicles. At present, the range of materials adaptive for FDM remains to be further studied, so as to accommodate the application of industrial mass production. It is imperative to control the machining parameters to decrease the surface porosity of the components. The support structure should also be subtly designed to minimize the impact on the quality of parts.

Aircraft Components [101] Additive manufacturing furnishes a prospective opportunity for multifunctional or integrated aircraft components. The jointless integrate aircraft components are intended to take the place of connectors, which wreck the structural integrity [102]. The prospective functionally graded materials offer customized material response and superior property to thermal environment or mechanical load, which will be a rising material for aerospace over the coming decades [103].

Energy Consumption and Savings The design for components is endowed with incomparable liberty from 3D printing technology. In particular, 3D printing technology has the capacity to fabricate light components and structure complex geometry, which can give rise to lessen the demand for energy and resources [2]. 3D printing technology can economize bunkers by means of cutting down the materials employed to manufacture aerospace components. Regarding lightweight as the basic argument, the mass decrease of 3D printed components signifies the fuel is depleted less. In a characteristic 30-year service life, for every 100 kg decreasing in aircraft mass, approximately 13.4–20 TJ of energy can be economized, which dramatically reduces the high energy consumption in the manufacturing stage [104]. Abating fuel consumption is momentous for the aviation industry.

In view of the five gradations of the 3D printing life cycle shown in Fig. 7, most of the recent researches on energy consumption of the 3D printing process have focused on the processing and manufacturing gradation and printing cycle graduation, namely, the second and third gradation [105]. Nevertheless, the best energy-saving gradation is estimated for applicational use, namely, the fourth gradation.

To establish several components in an integral 3D printing life cycle is aimed to lessen energy consumption. For instance, in the PBF process, the unmelted powder can be employed as a support for multiple parts within an integral 3D printing established volume, which can lessen the accumulative energy consumption by shortening the idle time of the entire printing process. Consequently, by attenuating the dependence on large convergent factories and assembly lines, the overall operational costs are further saved. Nevertheless, as a result of the excessive energy

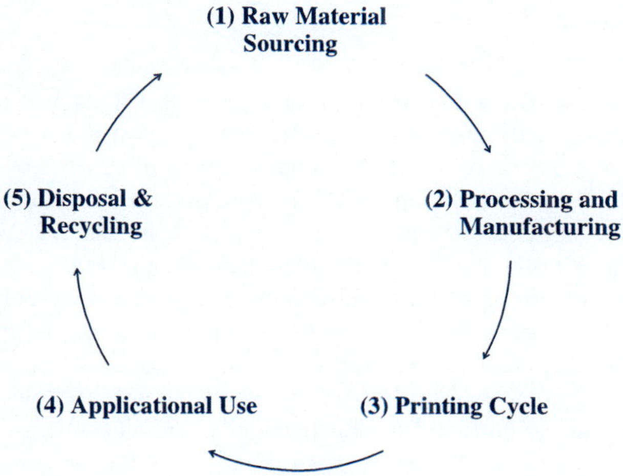

Fig. 7 Gradations of 3D printing life cycle. (Reproduced from [105])

consumption of the 3D printing process of metal powder, such energy saving is only able to be realized in the process of small batch production [106].

Multi-material Structures With the continuous increase of human demands, multi-material additive manufacturing has gradually appeared on the horizon of people. It combines with the advantages of a variety of materials, such as metals [107], ceramics, and polymers [108], to enhance the overall performance of one of them. For example, the combination of metal and ceramic materials increases wear resistance and corrosion resistance. The performance of the metal can also be improved by adding different phases to the new structure [109]. Therefore, multi-material additive manufacturing will revolutionize the aerospace industry.

In Situ Resource Utilization To authentically realize space exploration, we do with innovation. Exactly, space manufacturing has been promoting these innovations. Innovation is bound to change economics. NASA has been seeking to employ in situ resources for in-space manufacturing, which can help astronauts utilize materials at their disposal to make the equipment they demand. This will bring great progradation to the development of the aerospace industry. Mars is known as the earth's "sister star," as its topography, involving temperature and volume, is extremely close to earth year-round. Although Mars has been explored a lot, human understanding of Mars is not comprehensive. The application of 3D printing technology in Mars exploration will be a great prospect. The mystery of the universe explored constantly by human beings is unfathomable. Recently, a mass of researches has focused on building additive manufacturing processes to fabricate various desired materials utilizing in situ resources on Mars. This is a huge milestone.

8 Conclusions

In this review, additive manufacturing technology in the domain of aerospace is summarized. From the acquisition of raw materials, processing, and manufacturing to the diverse applications of aerospace, every gradation of additive manufacturing life cycle is closely linked and indispensable. Additive manufacturing technology has aroused the vigorous development of aerospace industry. New materials, such as intelligent materials, nanocomposites, and other materials with excellent comprehensive properties, are constantly explored to meet the demands of the development of aerospace industry. The range of applications is incrementally expanded, from accessories for commercial aircraft to space robots, satellites, and spaceships.

The most significant thing is that additive manufacturing enables the exploration journey of humans chasing the space dream further, and human cognition of the universe is becoming more distinct. People have gradually discovered a great many mysteries about the moon, Mars, and other planets, which are also inextricably linked with the earth. Additive manufacturing will be the mainstay of future aviation development. Simultaneously, we are also facing more technical difficulties, but only by breaking through them, we can really give full play to the best strength of 3D printing. Let's take the time to uncover the answer.

Acknowledgments This paper is supported by the Shanghai Sailing Program (Grant No. 19YF1434300), the National Natural Science Foundation of China (Grant No. 11947137), and the Shanghai Engineering Research Center of High-Performance Medical Device Materials (No. 20DZ2255500).

References

1. Duda, T., Raghavan, L.V.: 3D metal printing technology. IFAC-PapersOnLine. **49**(29), 103–110 (2016)
2. Joshi, S.C., Sheikh, A.A.: 3D printing in aerospace and its long-term sustainability. Virtual Phys. Prototyp. **10**(4), 175–185 (2015)
3. Oropallo, W., Ten Piegl, L.A.: Challenges in 3D Printing. Springer, Cham (2016)
4. Liu, Z., et al.: A critical review of fused deposition modeling 3D printing technology in manufacturing polylactic acid parts. Int. J. Adv. Manuf. Technol. **102**, 2877–2889 (2019)
5. Frazier, W.E.: Metal additive manufacturing: a review. J. Mater. Eng. Perform. **23**(6), 1917–1928 (2014)
6. Tofail, S.A.M., et al.: Additive manufacturing: scientific and technological challenges, market uptake and opportunities. Mater. Today. **21**(1), 22–37 (2017)
7. Erika, F., et al.: 3D printing of conductive complex structures with in situ generation of silver nanoparticles. Adv. Mater. **8**(19), 3712–3717 (2016)
8. Kruth, J.P., et al.: Progress in additive manufacturing and rapid prototyping. CIRP Ann. **47**(2), 525–540 (1998)
9. Sachs, E.M., et al.: Three dimensional printing: rapid tooling and prototypes directly from a CAD model. J. Eng. Ind. **39**(1), 201–204 (1992)
10. Skylar, T.: 4D printing: multi-material shape change. Archit. Des. **84**(1), 116–121 (2014)

11. Chua, C.K., Leong, K.F.: 3D Printing and Additive Manufacturing: Principles and Applications (with Companion Media Pack) – Fourth Edition of Rapid Prototyping. SG World Scientific Publishing Company, Singapore (2014)
12. Holmström, J., et al.: Rapid manufacturing in the spare parts supply chain. J. Manuf. Technol. Manag. **21**(6), 687–697 (2010)
13. Appleyard, M.: Corporate responses to online music piracy: strategic lessons for the challenge of additive manufacturing. Bus. Horiz. **58**(1), 69–76 (2015)
14. Weller, C., et al.: Economic implications of 3D printing: market structure models in light of additive manufacturing revisited. Int. J. Prod. Econ. **164**(164), 43–56 (2015)
15. Hofmann, M.: 3D printing gets a boost and opportunities with polymer materials. ACS Macro Lett. **3**(4), 382–386 (2014)
16. Gross, B.C., et al.: Evaluation of 3D printing and its potential impact on biotechnology and the chemical sciences. Anal. Chem. **86**(7), 3240–3253 (2014)
17. Shepherd, J.N.H., et al.: 3D microperiodic hydrogel scaffolds for robust neuronal cultures. Adv. Funct. Mater. **21**(1), 47–54 (2011)
18. Kim, K., et al.: The influence of stereolithographic scaffold architecture and composition on osteogenic signal expression with rat bone marrow stromal cells. Biomaterials. **32**(15), 3750–3763 (2011)
19. Derby, B.: Printing and prototyping of tissues and scaffolds. Science. **338**(6109), 921–926 (2012)
20. Vincenza, D., et al.: 3D-printed devices for continuous-flow organic chemistry. Beilstein J. Org. Chem. **9**(1), 951–959 (2013)
21. Symes, M.D., et al.: Integrated 3D-printed reactionware for chemical synthesis and analysis. Nat. Chem. **4**(5), 349–354 (2012)
22. Kong, Y.L., et al.: 3D printed quantum dot light-emitting diodes. Nano Lett. **14**(12), 7017–7023 (2014)
23. Espalin, D., et al.: 3D Printing multifunctionality: structures with electronics. Int. J. Adv. Manuf. Technol. **72**(5), 963–978 (2014)
24. Lewis, J.A., Ahn, B.Y.: Device fabrication: three-dimensional printed electronics. Nature. **518**(7537), 42 (2015)
25. Dimitrov, D., et al.: Advances in three dimensional printing – state of the art and future perspectives. Rapid Prototyp. J. **12**(3), 136–147 (2006)
26. Kruth, J.P.: Material incress manufacturing by rapid prototyping techniques. Ann. CIRP. **40**(2), 603–614 (1991)
27. Holzmann, P., et al.: User entrepreneur business models in 3D printing. J. Manuf. Technol. Manag. **28**(1), 75–94 (2017)
28. Stanic, M., et al.: Colorimetric properties and stability of 3D prints. Rapid Prototyp. J. **18**(2), 120–128 (2012)
29. Ngo, T.D., et al.: Additive manufacturing (3D printing): a review of materials, methods, applications and challenges. Compos. Part B Eng. **143**, 172–196 (2018)
30. O'Donnell, J., et al.: All-printed smart structures: a viable option? In: Active & Passive Smart Structures & Integrated Systems (2014). https://doi.org/10.1117/12.2045284
31. Pei, E.: 4D printing – revolution or fad? Assem. Autom. **34**(2), 123–127 (2014)
32. Pei, E.: 4D printing: dawn of an emerging technology cycle. Assem. Autom. **34**(4), 310–314 (2014)
33. Vaezi, M., et al.: A review on 3D micro-additive manufacturing technologies. Int. J. Adv. Manuf. Technol. **67**(5), 1721–1754 (2013)
34. Ford, S., Despeisse, M.: Additive manufacturing and sustainability: an exploratory study of the advantages and challenges. J. Clean. Prod. **137**(Nov. 20), 1573–1587 (2016)
35. Farahani, R.D., et al.: Three-dimensional printing of multifunctional nanocomposites: manufacturing techniques and applications. Adv. Mater. **28**(28), 5794–5821 (2016)
36. Keleş, Ö., et al.: Effect of build orientation on the mechanical reliability of 3D printed ABS. Rapid Prototyp. J. **23**(2), 320–328 (2017)

37. O'Brien, M.: Existing standards as the framework to qualify additive manufacturing of metals. In: 2018 IEEE Aerospace Conference (2018). https://doi.org/10.1109/AERO.2018.8396660
38. Singh, T., et al.: 3D printing of engineering materials: a state of the art review. Mater. Today Proc. **28**, 1927–1931 (2020)
39. Wang, Y., et al.: Selection of additive manufacturing processes. Rapid Prototyp. J. **23**(2), 434–447 (2017)
40. Singh, T., et al.: 3D printing of engineering materials: a state of the art review. Mater. Today Proc. **28**, 1927–1931 (2020)
41. Feldmann, M., et al.: Electromagnetic micro-actuators, micro-motors, and micro-robots. In: Microelectronics: Design, Technology, & Packaging III. Bellingham, Washington, DC (2007)
42. Gebisa, A.W., Lemu, H.G.: Investigating effects of fused-deposition modeling (FDM) processing parameters on flexural properties of ULTEM 9085 using designed experiment. Materials. **11**(4), 500 (2018)
43. Matsuzaki, R., et al.: Three-dimensional printing of continuous-fiber composites by in-nozzle impregnation. Sci. Rep. **6**, 23058 (2016)
44. Zaman, U.K.U., et al.: Impact of fused deposition modeling (FDM) process parameters on strength of built parts using Taguchi's design of experiments. Int. J. Adv. Manuf. Technol. **101**(5–8), 1215–1226 (2019)
45. Manakari, V., et al.: Selective laser melting of magnesium and magnesium alloy powders: a review. Metals. **7**(1), 35 (2017)
46. Sing, S.L., et al.: Characterization of titanium lattice structures fabricated by selective laser melting using an adapted compressive test method. Exp. Mech. **56**(5), 735–748 (2016)
47. Spierings, A.B., et al.: SLM-processed Sc- and Zr- modified Al-mg alloy: mechanical properties and microstructural effects of heat treatment. Mater. Sci. Eng. A. **701**, 264–273 (2017)
48. Shipley, H., et al.: Optimisation of process parameters to address fundamental challenges during selective laser melting of Ti-6Al-4V: a review. Int J Mach Tool Manu. **128**, 1–20 (2018)
49. Yu, W.H., et al.: Particle-reinforced metal matrix nanocomposites fabricated by selective laser melting: a state of the art review. Prog. Mater. Sci. **104**, 330–379 (2019)
50. Yuan, S., et al.: Polymeric composites for powder-based additive manufacturing: materials and applications. Prog. Polym. Sci. **91**, 141–168 (2019)
51. Hopkinson, N., et al.: Rapid Manufacturing: An Industrial Revolution for the Digital Age. Wiley, New York (2006)
52. Campbell, F.C.: Manufacturing Technology for Aerospace Structural Materials. Elsevier, Amsterdam (2006)
53. Uriondo, A., et al.: The present and future of additive manufacturing in the aerospace sector: a review of important aspects. Proc. Inst. Mech. Eng. Part G J. Aerosp. Eng. **229**(11), 0954410014568797 (2015)
54. Jia, Q., Gu, D.: Selective laser melting additive manufacturing of Inconel 718 superalloy parts: densification, microstructure and properties. J. Alloys Compd. **585**, 713–721 (2014)
55. Murr, L.E., et al.: Characterization of titanium aluminide alloy components fabricated by additive manufacturing using electron beam melting. Acta Mater. **58**(5), 1887–1894 (2010)
56. Mehrpouya, M., et al.: A prediction model for finding the optimal laser parameters in additive manufacturing of NiTi shape memory alloy. Int. J. Adv. Manuf. Technol. **105**(11), 4691–4699 (2019)
57. Rawal, S., et al.: Additive manufacturing of Ti-6Al-4V alloy components for spacecraft applications. 2013 6th International Conference on Recent Advances in Space Technologies (RAST) (2013)
58. Wu, X., et al.: Microstructures of laser-deposited Ti–6Al–4V. Mater. Des. **25**, 137–144 (2004)
59. Vilaro, T., et al.: As-fabricated and heat-treated microstructures of the Ti-6Al-4V alloy processed by selective laser melting. Metall. Mater. Trans. A. **42**(10), 3190–3199 (2011)
60. Romero, C., et al.: Fatigue and fracture properties of Ti alloys from powder-based processes – a review. Int. J. Fatigue. **117**(Dec.), 407–419 (2018)

61. Kahlin, M., et al.: Fatigue behaviour of notched additive manufactured Ti6Al4V with as-built surfaces. Int. J. Fatigue. **101**, 51–60 (2017)
62. Ning, F.D., et al.: Additive manufacturing of carbon fiber reinforced thermoplastic composites using fused deposition modeling. Compos. Part B Eng. **80**, 369–378 (2015)
63. Goh, G.D., et al.: Characterization of mechanical properties and fracture mode of additively manufactured carbon fiber and glass fiber reinforced thermoplastics. Mater. Des. **137**, 79–89 (2018)
64. Qiao, J., et al.: Ultrasound-assisted 3D printing of continuous fiber-reinforced thermoplastic (FRTP) composites. Addit. Manuf. **30**, 11 (2019)
65. Luo, M., et al.: Bi-scale interfacial bond behaviors of CCF/PEEK composites by plasma-laser cooperatively assisted 3D printing process. Compos. A: Appl. Sci. Manuf. **131**, 105812 (2020)
66. Anguita, J.V., et al.: Dimensionally and environmentally ultra-stable polymer composites reinforced with carbon fibres. Nat. Mater. **19**(3), 317–322 (2020)
67. Doreau, F., et al.: Stereolithography for manufacturing ceramic parts. Adv. Eng. Mater. **2**(8), 493–496 (2000)
68. Owen, D., et al.: 3D printing of ceramic components using a customized 3D ceramic printer. Prog. Addit. Manuf. **3**(1), 3–9 (2018)
69. Goulas, A., et al.: Additive manufacturing of physical assets by using ceramic multicomponent extra-terrestrial materials. Addit. Manuf. **10**, 36–42 (2016)
70. Karl, D., et al.: Towards the colonization of Mars by in-situ resource utilization: slip cast ceramics from Martian soil simulant. PLoS One. **13**(10), 11 (2018)
71. Faes, M., et al.: Extrusion-based additive manufacturing of ZrO2 using photoinitiated polymerization. CIRP J. Manuf. Sci. Technol. **14**, 28–34 (2016)
72. Ravindrababu, S., et al.: Evaluation of the influence of build and print orientations of unmanned aerial vehicle parts fabricated using fused deposition modeling process. J. Manuf. Process. **34**, 659–666 (2018)
73. Azarov, A.V., et al.: Composite 3D printing for the small size unmanned aerial vehicle structure. Compos. Part B Eng. **169**, 157–163 (2019)
74. None: 3D printed rocket thruster test success. Met. Powder Rep. **69**(4), 40 (2014). https://doi.org/10.1016/S0026-0657(14)70182-1
75. Wimpenny, D.I., et al.: Current trends of additive manufacturing in the aerospace industry (Chapter 4), pp. 39–54. https://doi.org/10.1007/978-981-10-0812-2_4 (2017)
76. None: 3D printed copper rocket engine part on way to Mars. Met. Powder Rep. **70**(4), 196–197 (2015). https://doi.org/10.1016/j.mprp.2015.06.021
77. Dhital, D., Ziegler, Y.: Additive manufacturing – application opportunities for the aviation industry. J. Air Transp. Stud. **6**, 63–86 (2015)
78. Schiller, G.J.: Additive manufacturing for aerospace. 2015 IEEE Aerospace Conference, pp. 1–8 (2015)
79. Tadjdeh, Y.: 3D printing promises to revolutionize defense, aerospace industries. Natl Def. **98**(724), 20–23 (2014)
80. Williamson, M.: Building a rocket? Press 'P' for print. Eng. Technol. **10**(2), 40–43 (2015)
81. Rawal, S. et al.: Additive manufacturing of Ti-6Al-4V alloy components for spacecraft applications. International Conference on Recent Advances in Space Technologies, pp. 5–112013
82. Kumar, L.J., et al.: Current trends of additive manufacturing in the aerospace industry. In: Advances in 3D Printing & Additive Manufacturing Technologies. Springer, Singapore (2017)
83. Matthews, N.: Chapter fifteen – Additive metal technologies for aerospace sustainment. In: Aircraft Sustainment & Repair, pp. 845–862. Butterworth-Heinemann, Oxford (2018)
84. Denkena, B., et al.: Engine blade regeneration: a literature review on common technologies in terms of machining. Int. J. Adv. Manuf. Technol. **81**(5–8), 917–924 (2015)
85. Halloran, J.W., et al.: Photopolymerization of powder suspensions for shaping ceramics. J. Eur. Ceram. Soc. **31**(14), 2613–2619 (2011)

86. Wang, Y.-C., et al.: Advanced 3D printing technologies for the aircraft industry: a fuzzy systematic approach for assessing the critical factors. Int. J. Adv. Manuf. Technol. **105**(10), 4059–4069 (2019)
87. Kohtala, C.: Addressing sustainability in research on distributed production: an integrated literature review. J. Clean. Prod. **106**(Nov. 1), 654–668 (2015)
88. Goh, G.D., et al.: A review on machine learning in 3D printing: applications, potential, and challenges. Artif. Intell. Rev. **54**(1), 63–94 (2021)
89. Zhang, Z.F., et al.: Audible sound-based intelligent evaluation for aluminum alloy in robotic pulsed GTAW: mechanism, feature selection, and defect detection. IEEE Trans. Ind. Inform. **14**(7), 2973–2983 (2018)
90. Okaro, I.A., et al.: Automatic fault detection for laser powder-bed fusion using semi-supervised machine learning. Addit. Manuf. **27**, 42–53 (2019)
91. Li, Z.X., et al.: Prediction of surface roughness in extrusion-based additive manufacturing with machine learning. Robot. Comput. Integr. Manuf. **57**, 488–495 (2019)
92. Mukherjee, T., Debroy, T.: A digital twin for rapid qualification of 3D printed metallic components. Appl. Mater. Today. **14**, 59–65 (2019)
93. Huang, S.H., et al.: Additive manufacturing and its societal impact: a literature review. Int. J. Adv. Manuf. Technol. **67**(5), 1191–1203 (2013)
94. El Moumen, A., et al.: Additive manufacturing of polymer composites: processing and modeling approaches. Compos. Part B Eng. **171**, 166–182 (2019)
95. Klippstein, H., et al.: Fused deposition modeling for unmanned aerial vehicles (UAVs): a review. Adv. Eng. Mater. **20**(2), 17 (2018)
96. Giri, J., et al.: Optimization of FDM process parameters for dual extruder 3d printer using artificial neural network. Mater. Today Proc. **43**, 3242–3249 (2021)
97. Witze, A.: NASA to send 3D printer into space. Nature. **513**(7517), 156 (2014)
98. Guanghui, S., et al.: An aerospace bracket designed by thermo-elastic topology optimization and manufactured by additive manufacturing. Chin. J. Aeronaut. **33**(4), 1252–1259 (2020)
99. DebRoy, T., et al.: Additive manufacturing of metallic components – process, structure and properties. Prog. Mater. Sci. **92**, 112–224 (2018)
100. Du Plessis, A., Macdonald, E.J.A.M.: Hot isostatic pressing in metal additive manufacturing: X-ray tomography reveals details of pore closure. Addit. Manuf. **34**, 101191 (2020)
101. Liu, R.: Aerospace applications of laser additive manufacturing. In: Laser Additive Manufacturing (2017). https://doi.org/10.1016/B978-0-08-100433-3.00013-0
102. Gausemeier, J., et al.: Thinking Ahead the Future of Additive Manufacturing – Scenario-Based Matching of Technology Push and Market Pull. RTejournal - Forum fiir Rapid Technologie (2012). https://rtejournal.de/paper
103. Cooley, W.G.: Application of Functionally Graded Materials in Aircraft Structures. AIR FORCE INSTITUTE OF TECHNOLOGT (2012). https://www.docin.com/p-1766849197.html
104. Huang, R., et al.: Energy and emissions saving potential of additive manufacturing: the case of lightweight aircraft components. J. Clean. Prod. **135**(Nov. 1), 1559–1570 (2016)
105. Serres, N., et al.: Environmental comparison of MESO-CLAD® process and conventional machining implementing life cycle assessment. J. Clean. Prod. **19**(9), 1117–1124 (2011)
106. Singamneni, S., et al.: Additive Manufacturing for the Aircraft Industry: A Review. journal of aeronautics and aerospace engineering. **8**(1), 1–13 (2019)
107. Hoefer, K., et al.: Multi-material additive manufacturing by 3D plasma metal deposition for graded structures of super duplex alloy 1.4410 and the austenitic corrosion resistant alloy 1.4404. JOM. **71**(4), 1554–1559 (2019)
108. Singh, R., et al.: Multi-material additive manufacturing of sustainable innovative materials and structures. Polymers. **11**(1), 14 (2019)
109. Bandyopadhyay, A., Heer, B.: Additive manufacturing of multi-material structures. Mater. Sci. Eng. R Rep. **129**, 1–16 (2018)

Instantaneous Availability Analysis of Maintenance Process Based on Semi-Markov Model

Yi Yang, Tingting Zeng, Siyu Huang, and Wei Liu

1 Introduction

In the using process of equipment, availability describes the ability of equipment to perform the specified tasks normally at any time. It can also measure operational readiness and mission sustainability. Current researches on availability mainly focus on steady-state availability application and instantaneous availability modeling in complex systems. Steady-state availability reflects the usability when the system reaches a stable operation after a long running time. However, in the early stage of equipment use, it fails to describe both the fluctuation of operational capability and the variety of availability. In addition, the analysis of the initial fluctuation of equipment availability can effectively reflect the real-time performance of equipment.

There are three main factors that affect the availability of equipment, namely, reliability, maintainability, and supportability [1–4]. The influence of reliability and maintainability can be reduced by reasonable design of development period. However, if the system fails to attain support resource after failure, it will delay the maintenance action, induce support delay, and impact the availability. The concept of support delay was presented in 1973 [5]. As the lucubration of support delay on system availability, support delay time comes into focus [6, 7]. Reference [8] utilized the Markov process and Laplace transform to obtain an expression for steady-state availability after establishing a model that contains the time of prevention and maintenance. Reference [9] used the Markov renewal process theory and total probability decomposition technique to obtain the steady-state availability of the system. There are many available models that consider support delay, but

Y. Yang · T. Zeng · S. Huang (✉) · W. Liu
Beihang University, Beijing, China

D. Tang et al. (eds.), *Mobile Wireless Middleware, Operating Systems and Applications*, EAI/Springer Innovations in Communication and Computing,
https://doi.org/10.1007/978-3-030-98671-1_7

majority of them concentrate on steady-state availability. Meanwhile, researches of support delay time usually center on causing by a single support resource, such as equipment delay, queue waiting, and others [10, 11]. The study caused by multiple support factors is rarely being launched. Therefore, this paper will comprehensively consider the impact of multiple factors on instantaneous availability of support process.

Support delay is mainly caused by three factors: equipment, spares, and personnel. The system has several possible states while waiting for maintenance. The available modeling methods of multistate system are divided into four categories, namely, Boolean extended model method, Markov model method, general generation function method, and Monte Carlo simulation method. Markov model method can effectively characterize the state transformation process of the system and deconstruct transition speciality of multistate. Reference [12] introduced a special method called LZ transformation to analyze instantaneous availability in Markov process of discrete state and continuous time. During support process, the state stationary time does not fully satisfy exponential distribution, which is a condition of standard Markov process. For this situation, researchers use simulation methods to study it [13]. In Reference [14], an analytical and probabilistic availability model for periodical inspection system is proposed by a new recursive algorithm, which can achieve limiting average availability and instantaneous availability of periodical inspection system under arbitrary lifetime and repair-time distributions. In Reference [15–17], by using traditional empirical modeling methods, they established availability evaluation models of complex systems. However, facing the transition process between different states, traditional model is unsuitable. Meanwhile, it is more universal to implement semi-Markov process in the modeling of support delay process that obeys random distribution. At present, researches on multistate system reliability modeling with semi-Markov transition are scarce [18–20]. So the application of support delay construction remains to be studied.

In this paper, based on semi-Markov method, we established a more universal instantaneous availability analysis model, which, considering delay factors of support personnel, supports equipment and spares. The analytic expression of instantaneous availability is obtained by Fourier transform. Besides, we investigated the change of instantaneous availability over time.

The paper is organized as follows. In Sect. 2, we present related definitions of semi-Markov model. In Sect. 3, the instantaneous availability model of repairable system considering support delay is proposed, and a numerical example is given in Sect. 4.

2 Preliminaries

The assumption that state performance space is discrete while time course space is continuous is more corresponding to engineering reality of state transition. So we only based on continuous semi-Markov model for mathematical interpretation.

We consider a system S With m possible states, m is a finite natural number. The set of all possible states is noted as $I = \{1, 2, \cdots, m\}$. For discrete time $n \in N_0$, $N_0 = \{0, 1, 2, \cdots\}$, the initial state of S at time 0 is represented by random variable J_0. The system S stays a random length of time X_1 in the initial state, then enters the next state J_1 to stay for a random length of time X_2 before going into J_2, and so on. For discrete time, the sequence $(J_n, n \geq 0)$ represents the successive states of S. The sequence $(X_n, n \geq 0)$ gives the successive sojourn time. The two-dimensional stochastic process in discrete time is called a positive $(J - X)$ process:

$$(J - X) = ((J_n - X_n), n \geq 0). \tag{1}$$

Supposing $X_0 = 0$, $a.\ s.$ The time of state transition is given as $(T_n, n \geq 0)$, where

$$T_n = \sum_{i=1}^{n} X_i. \tag{2}$$

Specially, for continuous time t, the state of system S is represented by r.v. $J(t)$. Let $J(t)$ be the semi-Markov process corresponding to the positive $(J - X)$ process. The time point is recorded as $t_1, t_2, \cdots, t_n$. The state sojourn time is recorded as x_n, $x_n = t_n - t_{n-1}$.

Based on the probability space, $P(J(0) = i) = p_i$, $i = 1, \cdots, m$ with $\sum_{i=1}^{m} p_i = 1$ is given. We assume that for all $n \in N_0, j \in I, t \in R^+$:

$$\begin{aligned} P\left(J\left(t_n\right) = j, x_n \leq t \mid J\left(t_{n-1}\right) = i, x_{n-1}, J\left(t_{n-1}\right), x_{n-2}, \cdots, J_0\right) \\ = P\left(J\left(t_n\right) = j, x_n \leq t \mid J\left(t_{n-1}\right) = i\right). \end{aligned} \tag{3}$$

We noted the running states of system S as

$$Q_{ij}(t) = P\left(J\left(t_n\right) = j, x_n \leq t \mid J\left(t_{n-1}\right) = i\right). \tag{4}$$

where $Q_{ij}(t)$, $i, j \in I$, $t \in [0, t_n - t_{n-1}]$ is a nondecreasing real function null on R^+ such that if

$$p_{ij} = P\left\{J\left(t_n\right) = j, \mid J\left(t_{n-1}\right) = i\right\} = \lim_{t \to \infty} Q_{ij}(t), i, j \in I, \tag{5}$$

then

$$\sum_{j=1}^{m} p_{ij} = 1, i \in I. \tag{6}$$

We can write the matrix: $\mathbf{Q} = [Q_{ij}]$, $\mathbf{P} = [p_{ij}]$.

Definition 2.1. [21] Every $m \times m$ matrix $\mathbf{Q}$ of nondecreasing functions null on R^+ satisfying properties Eqs. 5 and 6 is called a semi-Markov matrix or a semi-Markov kernel.

Definition 2.2. [21] Every couple $(\mathbf{P}, \mathbf{Q})$ where $\mathbf{Q}$ is a semi-Markov kernel and $\mathbf{P}$ a vector of initial probabilities defines a positive $(J - X)$ process as state space and is also called a semi-Markov chain.

The conditional sojourn time distribution function in state i, given that the next state is j, is F_{ij}. The corresponding probability density function is f_{ij}.

We have

$$F_{ij}(t) = P\left(x_n \leqslant t | \; J(t_n) = j, J(t_{n-1}) = i\right) = \int_{-\infty}^{t} f_{ij}(u) = \begin{cases} \dfrac{Q_{ij}(t)}{p_{ij}} & \text{when } p_{ij} > 0 \\ 1 & \text{when } p_{ij} = 0 \end{cases} \tag{7}$$

The unconditional distribution function of sojourn time in state i can be denoted as

$$F_i(t) = P\left\{x_n \leqslant t | \; J(t_{n-1}) = i\right\} = \sum_{j \in I} Q_{ij}(t) \tag{8}$$

For the calculation of $Q_{ij}(t)$, Reference [22] gives an integral decomposition algorithm, which can be applied to any countable state transitions.

Assuming $F_{ij}(t)$ is known, and the system is currently in j state, it may shift to p state, q state, or o state at next moment. Then we have:

$$\begin{cases} Q_{jp}(t) = \int_0^t \left(1 - F_{jq}(u)\right)\left(1 - F_{jo}(u)\right) dF_{jp}(u) \\ Q_{jq}(t) = \int_0^t \left(1 - F_{jp}(u)\right)\left(1 - F_{jo}(u)\right) dF_{jq}(u) \\ Q_{jo}(t) = \int_0^t \left(1 - F_{jp}(u)\right)\left(1 - F_{jq}(u)\right) dF_{jo}(u) \end{cases} . \tag{9}$$

3 Instantaneous Availability Analysis of Maintenance Process

Support process is a critical link to ensure system availability. The failure modes of an advanced material are unclear, so the supply capacity of spare storage is very weak. In the using process, materials may not get repairs in time after shutting down with a malfunction. Majority commit waits for maintenance; hence, support delay occurs. The three-state situation including work, support delay, and maintenance compared to two-status condition is more corresponding to actuality [23]. There are abundant reasons for support delay such as management, transportation, personnel,

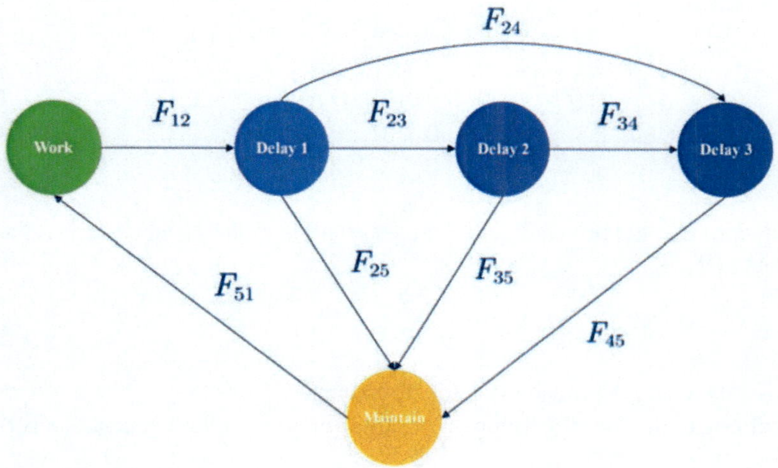

Fig. 1 Sketch map of conditional sojourn time for the semi-Markov process

equipment, spares and storage, etc. We summarized the main factors that cause support delay into three aspects, namely, cover personnel delay time, equipment delay time, and spares delay time.

Based on the above analysis, this section established an instantaneous availability model that considers support delay for repairable system. The possible states of materials are as follows.

State 1: Work. The system is working normally.
State 2: Delay. 1. In the process of support delays, the system is waiting for support personnel to detect after failing.
State 3: Delay. 2. In the process of support delays, the system is waiting for the equipment after failing while support personnel are in place.
State 4: Delay. 3. In the process of support delays, the system is waiting for spares while personnel and equipment are in place.
State 5: Maintain. The system is repairing while support resources are all in place.

Maintenance process of system after failure is shown in Fig. 1.

The support delay process can be regarded as a system with five states. The set of all possible states is noted as $I = \{1, 2, 3, 4, 5\}$.

To keep the system in normal working condition during operating, maintenance and support progress are significant links. The state space $U (U \subseteq I)$ represents normal working state of system where instantaneous availability is the probability of the system that is in the working state:

$$A(t) = P\left(J(t) \in U\right). \tag{10}$$

We consider the following semi-Markov kernel matrix Q:

$$Q = \begin{bmatrix} 0 & Q_{12} & 0 & 0 & 0 \\ 0 & 0 & Q_{23} & Q_{24} & Q_{25} \\ 0 & 0 & 0 & Q_{34} & Q_{35} \\ 0 & 0 & 0 & 0 & Q_{45} \\ Q_{51} & 0 & 0 & 0 & 0 \end{bmatrix}. \tag{11}$$

We record that the probability when the system is in the j state at t triggered by the i state is ψ_{ij}:

$$\psi_{ij}(t) = P\{J(t) = j | J(0) = i\}, \tag{12}$$

which is called trigger probability in the following.

According to the transition process of semi-Markov, solution conclusion is [21]:

$$\psi_{ij}(t) = \delta_{ij}(1 - F_i(t)) + \sum_{r=1}^{5} \int_0^t \sigma_{ir}(\tau)\psi_{rj}(t-\tau)d\tau, \tag{13}$$

where

$$\delta_{ij} = \begin{cases} 1, i = j \\ 0, i \neq j \end{cases}, \sigma_{ir}(\tau) = \frac{dQ_{ir}(\tau)}{d\tau}. \tag{14}$$

The instantaneous availability can be calculated as follows:

$$A(t) = \sum_{i \in I} \sum_{j \in U} p_i \psi_{ij}(t). \tag{15}$$

4 Analysis Case of Instantaneous Availability

We study system instantaneous availability in the situation that considers support delay and supposes the system initial state is in working. Hence, the initial probability vector is

$$\boldsymbol{p} = (p_1, p_2, p_3, p_4, p_5) = (1, 0, 0, 0, 0). \tag{16}$$

According to maintenance process experiments of an aeronautic electronic component, the conditional sojourn time distribution can be obtained (Table 1).

According to Eq. 9, solve the element Q_{ij} in the semi-Markov kernel matrix Q:

$$Q_{12}(t) = \int_0^t f_{12}(u)du = F_{12}(t) = 1 - e^{-(t/500)^{1.5}} \tag{17}$$

Table 1 The distribution function of sojourn time in state transition

State i	State j	F_{ij}	Parameter
1	2	Weibull distribution $1 - e^{-(t/\lambda)^k}$	$\lambda = 500$ $k = 1.5$
2	3	Normal distribution $\int_0^t \frac{1}{\sqrt{2\pi}\sigma} \exp\left(\frac{(u-\mu)^2}{2\sigma^2}\right) du$	$\mu = 3$ $\sigma^2 = 1$
2	4	Normal distribution $\int_0^t \frac{1}{\sqrt{2\pi}\sigma} \exp\left(\frac{(u-\mu)^2}{2\sigma^2}\right) du$	$\mu = 10$ $\sigma^2 = 2$
2	5	Exponential distribution $1 - e^{-\lambda t}$	$\lambda = 0.05$
3	4	Normal distribution $\int_0^t \frac{1}{\sqrt{2\pi}\sigma} \exp\left(\frac{(u-\mu)^2}{2\sigma^2}\right) du$	$\mu = 7$ $\sigma^2 = 2$
3	5	Exponential distribution $1 - e^{-\lambda t}$	$\lambda = 0.05$
4	5	Exponential distribution $1 - e^{-\lambda t}$	$\lambda = 0.05$
5	1	Normal distribution $\int_0^t \frac{1}{\sqrt{2\pi}\sigma} \exp\left(\frac{(u-\mu)^2}{2\sigma^2}\right) du$	$\mu = 20$ $\sigma^2 = 10$

$$\begin{cases} Q_{23}(t) = \int_0^t (1 - F_{24}(u))\,(1 - F_{25}(u))\,dF_{23}(u) \\ Q_{24}(t) = \int_0^t (1 - F_{23}(u))\,(1 - F_{25}(u))\,dF_{24}(u) \\ Q_{25}(t) = \int_0^t (1 - F_{23}(u))\,(1 - F_{24}(u))\,dF_{25}(u) \end{cases} \tag{18}$$

$$\begin{cases} Q_{34}(t) = \int_0^t (1 - F_{35}(u))\,dF_{34}(u) \\ Q_{35}(t) = \int_0^t (1 - F_{34}(u))\,dF_{35}(u) \end{cases} \tag{19}$$

$$Q_{45}(t) = \int_0^t f_{45}(u)du = F_{45}(t) = 1 - e^{-0.05t} \tag{20}$$

$$Q_{51}(t) = \int_0^t f_{51}(u)du = F_{51}(t) = \int_0^t \frac{1}{\sqrt{200\pi}} \exp\left(\frac{(u-20)^2}{20}\right) du. \tag{21}$$

Because some analytical solutions cannot be obtained, the trapezoidal area method is used to solve the numerical solution as shown in Figs. 2, 3, 4, 5 and 6.

Further, according to Eq. 13, we can get ψ_{ij}. Eq. 13 is a system of 5*5 nonlinear integral equations with convolution operation. We use the special properties of convolution in Fourier transform, which is carried out on both sides of the equation group. Then there are

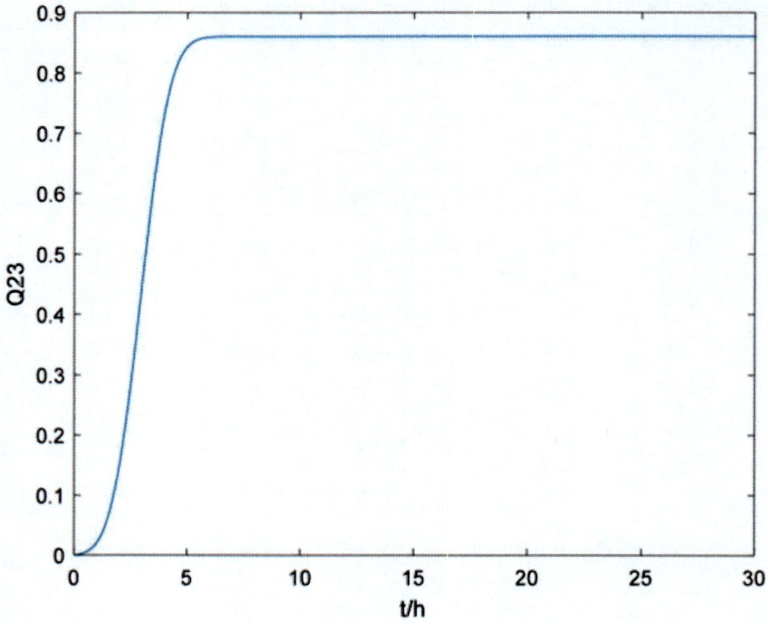

Fig. 2 Fluctuation graph of element Q_{23}

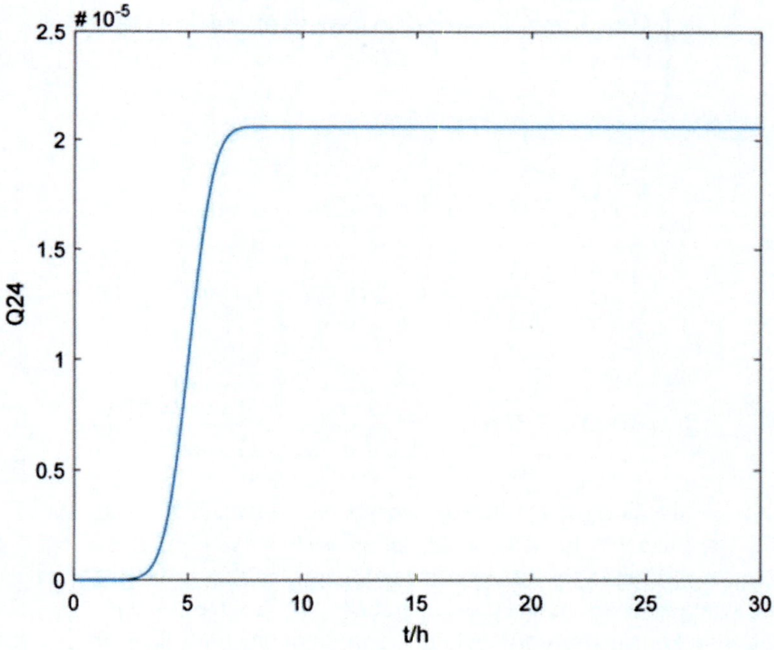

Fig. 3 Fluctuation graph of element Q_{24}

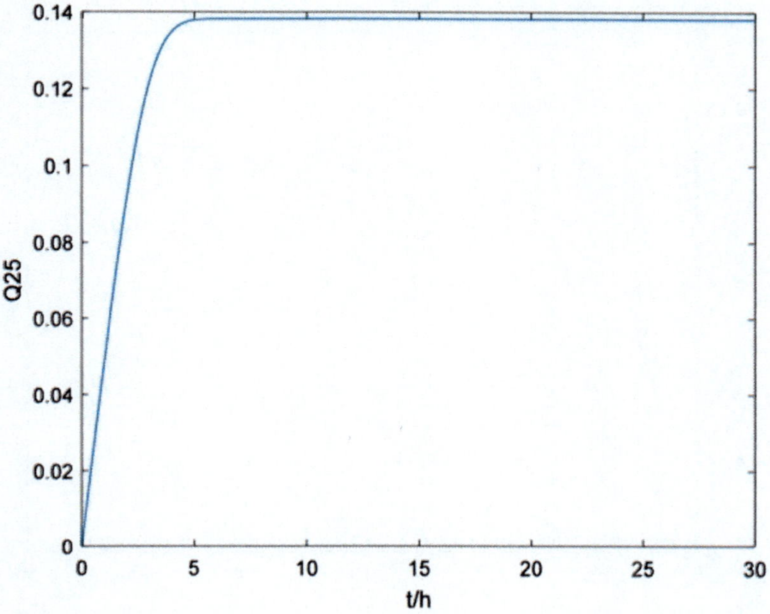

Fig. 4 Fluctuation graph of element Q_{25}

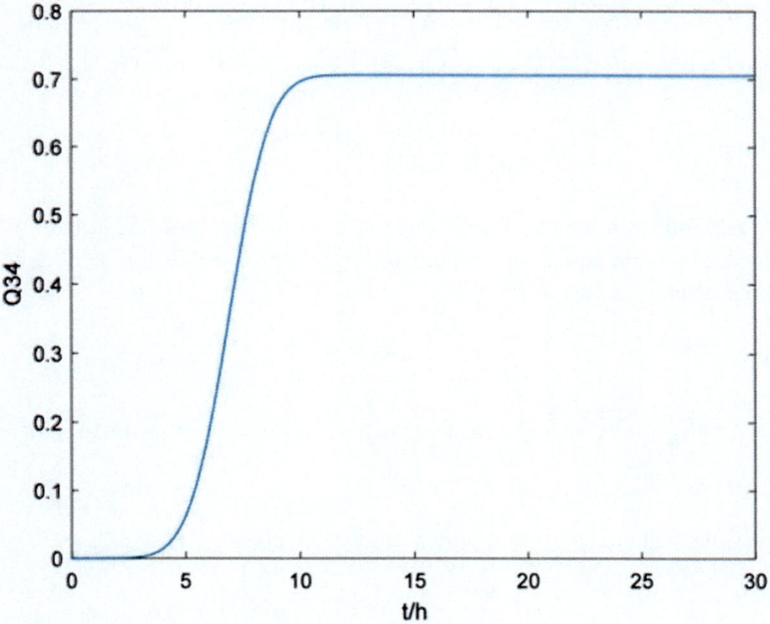

Fig. 5 Fluctuation graph of element Q_{34}

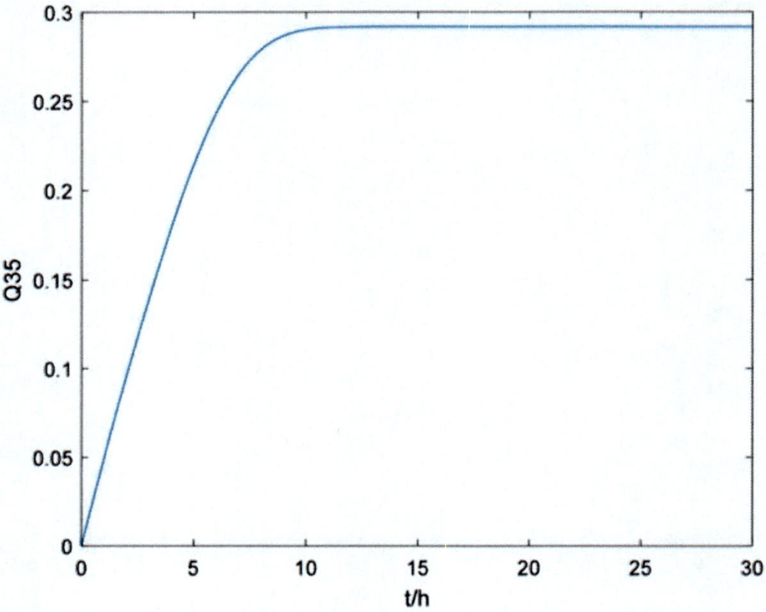

Fig. 6 Fluctuation graph of element Q_{35}

$$\begin{cases} \Upsilon[\psi_{11}] = \Upsilon[1 - F_1] + \Upsilon[\sigma_{12}]\Upsilon[\psi_{21}] \\ \Upsilon[\psi_{21}] = \Upsilon[\sigma_{23}]\Upsilon[\psi_{31}] + \Upsilon[\sigma_{24}]\Upsilon[\psi_{41}] + \Upsilon[\sigma_{25}]\Upsilon[\psi_{51}] \\ \Upsilon[\psi_{31}] = \Upsilon[\sigma_{34}]\Upsilon[\psi_{41}] + \Upsilon[\sigma_{35}]\Upsilon[\psi_{51}] \\ \Upsilon[\psi_{41}] = \Upsilon[\sigma_{45}]\Upsilon[\psi_{51}] \\ \Upsilon[\psi_{51}] = \Upsilon[\sigma_{51}]\Upsilon[\psi_{11}] \end{cases}. \tag{22}$$

Υ is the Fourier operator. This change is used to transform the equations into a nonlinear algebraic equations. Finally, the trigger probability can be obtained by using Fourier inverse operator:

$$\psi_{11} = \Upsilon^{-1} \left\{ \frac{\Upsilon[1 - F_1]}{1 - \Upsilon[\sigma_{12}]\Upsilon[\sigma_{23}]\Upsilon[\sigma_{51}] \cdot (\Upsilon[\sigma_{34}]\Upsilon[\sigma_{45}] + \Upsilon[\sigma_{35}]) - \Upsilon[\sigma_{12}]\Upsilon[\sigma_{51}] \cdot (\Upsilon[\sigma_{24}]\Upsilon[\sigma_{45}] + \Upsilon[\sigma_{25}])} \right\}. \tag{23}$$

Finally, the instantaneous availability can be calculated:

$$A(t) = \sum_{i \in I} \sum_{j \in U} p_i \psi_{ij}(t) = \psi_{11}(t), \tag{24}$$

and its image fluctuating with time is shown in Fig. 7.

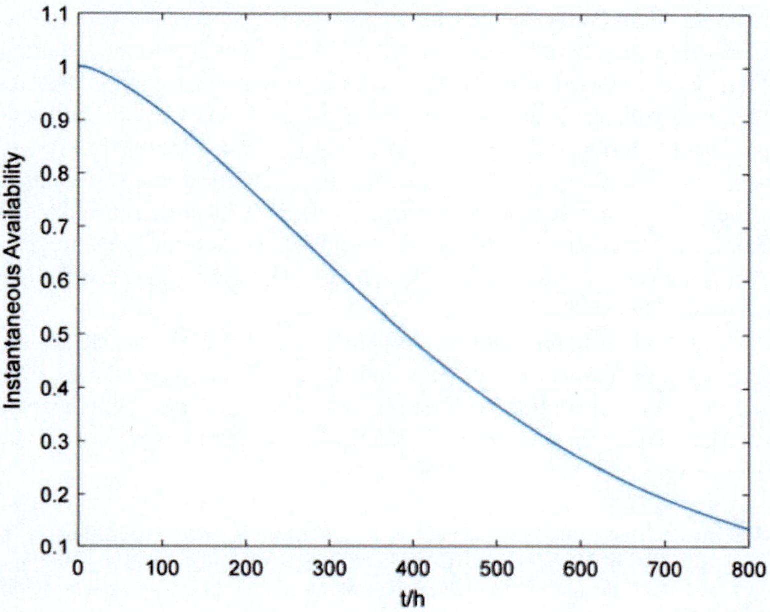

Fig. 7 The curve of instantaneous availability

According to Fig. 7, we can see that the instantaneous availability has an extremely fast drop in early stage, and then it is more stable for a short period of time. With the increase of time, system instantaneous availability decreases gradually. We consider that the system is in the early stage of failure; therefore, the increase of failure rate affects instantaneous availability. Subsequently, system performance is stable, and the availability decreases with time steadily. Finally, in the process of approaching the service life, system enters an unusable state.

5 Conclusion

In this paper, actual situation transforms of system in the process of maintenance and support are considered, and the delay phenomenon of support activities in system is analyzed. The traditional three-state systems with support delay are extended to multi-support factor simultaneous delay system. Based on the semi-Markov process, we first construct support delay model caused by the lack of multiple support resources, which describes the delayed state migration process of support personnel, support equipment, and spares in logistics support. The numerical expression of instantaneous availability with time is obtained by solving convolution equations in Fourier transform. The change of instantaneous availability with time is analyzed, and it is found that the support delay has a great influence on the availability of the system. In order to reduce its impact, we can increase the number of support personnel, improve the attendance rate of support personnel, shorten the time of arrival of support equipment, and reserve sufficient spares. The results enrich the instantaneous availability modeling of multistate system, promote the development of solution technology in semi-Markov transition environment, and implement real-time tracking of instantaneous availability numerical changes. It has considerable engineering value to promote synchronous precision and matching construction of related support resources. In this paper, the maintenance process model corresponds with actuality; therefore, we suppose that it is accessible to prove system maintenance process.

However, the three main influencing factors considered at present are macro. The specific segmentation remains to probe. It is essential to gradually improve maintenance process modeling with various elements that may influence system property. Meanwhile, the influence mechanism of different factors needs to be further explored.

Acknowledgments This work was supported by National Science Foundation of China under Grants 61871013, 71671009 and Consulting Project of China Academy of Engineering:2021XY39. Finally, the authors would like to thank the editors and anonymous reviewers for their insightful comments.

References

1. Yang, Y., Yu, Y.L., Wang, L.C.: Fluctuation Mechanism and Control on System Instantaneous Availability. CRC Press, Boca Raton (2015)
2. Yang, Y., Ren, S.C., Fan, G.M., Kang, R.: Numerical simulation on the existence of fluctuation of instantaneous availability. Trans. Can. Soc. Mech. Eng. **40**(5), 703–713 (2016)
3. Yang, Y., Chen, S.C., Yu, Y.L.: Theoretical analysis of instantaneous availability of systems under uniform distribution. J. Beihang Univ. **42**(001), 28–34 (2016)
4. Yang, J.K., Xu, T.X., Wang, H.W., Mi, Q.L.: Research on the method of modeling availability of missile weapon systems. Tactical Missile Technol. **05**, 30–34 (2012)
5. Wenyuan, L.U., Wang, W., Christer, A.H.: The delay time modeling of preventive maintenance of plant based on subjective PM data and actual failure records. In: International Conference on Quality & Reliability(ICQR2005) School of Management. University of Shanghai for Science & Technology, P O Box475, JunGong Road 516,Shanghai (200093), China (2005)
6. Wang, W.: A two-stage prognosis model in condition based maintenance. Eur. J. Oper. Res. **182**(3), 1177–1187 (2007)
7. Aven, T., Castro, I.T.: A delay-time model with safety constraint. Reliab. Eng. Syst. Saf. **94**(2), 261–267 (2009)
8. Liu, F.S., Wu, W., Shan, Z.W., Chen, Y.: Useability model of armored equipment based on Markov update process. J. Armoured Corps Eng. **24**(005), 19–21 (2010)
9. Wei, Y.Q., Tang, Y.H.: Cold storage and repair system for two different parts based on replacement and repair delay strategy of repair equipment. J. Eng. Math. **37**(04), 423–441 (2020)
10. Yang, Y., Chen, Y., Wen, M.L.: Analysis of instantaneous availability of communication system based on the influence of support equipment. Int. J. Commun. Syst. **31**(99), e3480 (2018)

11. Yongli, Y., Liu, Z.: Basic Theory and Method of Equipment Support Engineering. National Defense Industry Press, Beijing (2015)
12. Lisnianski, A., Frenkel, I., Karagrigoriou, A.: Recent Advances in Multi-State Systems Reliability: Theory and Applications. Springer, Cham (2018)
13. Ruan, Y.P.: Study on Reliability Evaluation Method of Complex System Based on Monte Carlo Simulation. Tianjin University, Tianjin (2013)
14. Li, J.L., Chen, Y.L., Zhang, Y., Huang, H.L.: Availability modeling for periodically inspection system with different lifetime and repair-time distribution. Chin. J. Aeronaut. **32**(7), 1667–1672 (2019)
15. Kong, D.Z., Li, X.B.: A instantaneous availability modelling method for repairable system. Appl. Mech. Mater. **3764**(724), 334–339 (2015)
16. Zhang, H., Meng, D.B., Zong, Y.Y., Wang, F., Xin, T.L.: A modeling and analysis strategy of constellation availability using on-orbit and ground added launch backup and its application in the reliability design for a remote sensing satellite. Adv. Mech. Eng. **10**(4), 1–6 (2018)
17. Fan, R.N.: Theory and Method of Transient Index Approximation in Repairable System. Beijing Institute of Technology, Beijing (2015)
18. Wang, L., Li, M.: Redundancy allocation optimization for multistate systems with failure interactions using semi-Markov process. J. Mech. Des. **137**(10), 101403 (2015)
19. Chryssaphinou, O., Limnios, N., Malefaki, S.: Multi-state reliability systems under discrete time semi-Markovian hypothesis. IEEE Trans. Reliab. **60**(1), 80–87 (2011)
20. Shang, Y.L., Cai, Q., Zhao, X.W., Chen, L.: Multi-state reliability analysis of reactor pump unit based on UGF and semi-Markov methods. Nucl. Power Eng. **33**(1), 117–123 (2012)
21. Janssen, J., Manca, R.: Applied Semi-Markov Processes. Springer, New York (2006)
22. Limnios, N., Opri, G.: An Semi-Markov Processes and Reliability. Birkhauser, Basel (2001)
23. Ren, S., Yang, Y., Chen, Y., Du, Y.: Fluctuation analysis of instantaneous availability under specific distribution. Neurocomputing. **270**, 152–158 (2017)

A Survey of Techniques for Constructing Mongolian Domain-Specific Knowledge Graph

Gegerihu Bao, Haishan Bao, Dalai Tang, Arong Suyila, and A. Gudamu

1 Introduction

In 2012, Google originally introduced the conception of "Knowledge Graph" on their official blog site. As described, *knowledge graph* can provide richer search capabilities for the search engines to improve search quality and user experience, making search about *things, not strings* [1]. *Knowledge graph* is a particular database that amalgamates information in a structured format that can explicitly represent the relations between entities. In brief, the knowledge graph is composed of some interconnected entities and their attributes. Each piece of knowledge in the knowledge graph can be represented as a triplet, which is a 3-tuple (s, r, e) where s and e, respectively, represent the start and end entities, and r represents a relationship between the two entities [2]. $G = (E, R, S)$ refers to the general expression of *knowledge graph*, where $E = \{e_1, e_2, e_3, \cdots, e_{|E|}\}$ denotes the entity

Supported by the Department of Finance and the Department of Science and Technology of Inner Mongolia Autonomous Region, China (Project No. 2019CG065), and the Department of Education of Inner Mongolia Autonomous Region, China (Project No. NJZY22265).

G. Bao
School of Mathematics and Big Data Science, Hohhot College for Nationalities, Hohhot, China
e-mail: gegerihu.bao@imnc.edu.cn

H. Bao (✉) · D. Tang
Inner Mongolia University of Finance and Economics, Hohhot, China
e-mail: bhs@imufe.edu.cn; tdl@imufe.edu.cn

A. Suyila · A. Gudamu
Inner Mongolia Dayaar Technology Co., Ltd, Inner Mongolia, China
e-mail: arongsuyila@dayaar.com.cn; agudamu@dayaar.com.cn
http://www.dayaar.com.cn/

D. Tang et al. (eds.), *Mobile Wireless Middleware, Operating Systems and Applications*, EAI/Springer Innovations in Communication and Computing,
https://doi.org/10.1007/978-3-030-98671-1_8

set, $R = \{r_1, r_2, r_3, \cdots, r_{|R|}\}$ denotes the relation set, and $S \subseteq E \times R \times E$ represents the set of all triples in the knowledge graph.

Nowadays, knowledge graph has become the cornerstone of intelligent applications, such as semantic search systems, knowledge-based conversion, decision-making, knowledge representation and reasoning, and recommendation systems. At present, researchers have constructed and released various kinds of knowledge graphs, such as Wikidata [3], YAGO [4], DBpedia [5], and Freebase [6]. The works mentioned above are all in English and related to the general domain. Recently, a lot of Chinese knowledge graphs have also been completed, such as CN-DBpedia that is a general domain knowledge graph developed by Fudan University [7], zhishi.me [8], which is also a general domain knowledge graph, and the data is extracted from Chinese encyclopedias such as Baidu Baike. Ownthink [9] opens up tools for conversational robots, knowledge mapping, semantic understanding, natural language processing, XLore [10], etc.

Mongolian traditional script is almost used by 6 million people in Inner Mongolia and other autonomous regions or provinces in China [11]. Mongolian people use it since the era of Genghis Khan around the twelfth century, which is still one of the large ethnic languages in China. Mongolian is an agglutinative language, which has a significantly different vertical writing model, multiple font type variations, and a great size of the vocabulary. All these features bring challenges to NLP studies of traditional Mongolian. The development of the Mongolian Unicode system is much slower than other languages. In Inner Mongolia, there are many kinds of code systems for Mongolian traditional scripts. Lots of Mongolian data sources are stored by different code systems. In recent years, under the support of government and industrial communities, the informatization progress of Mongolian language has gotten a great achievement.

2 Related Work

2.1 Construction of Knowledge Graph

In this section, we briefly introduce the process of constructing a knowledge graph. There are two types of architecture of knowledge graph construction, one is the top-down approach, and the other one is the bottom-up approach [12]. The top-down architecture means that the ontology and schema of knowledge already have been well-defined, and new knowledge instances can be directly added into the knowledge database. The bottom-up architectures is that knowledge is extracted from knowledge data resources. The top-down architecture suit for constructing a knowledge graph that knowledge instances are already stored as structured or in some kind of format. If not, the bottom-up approach is recommended.

The construction process of the knowledge graph can be divided into several parts. Many previous studies have shown about it [13, 14]. Although different

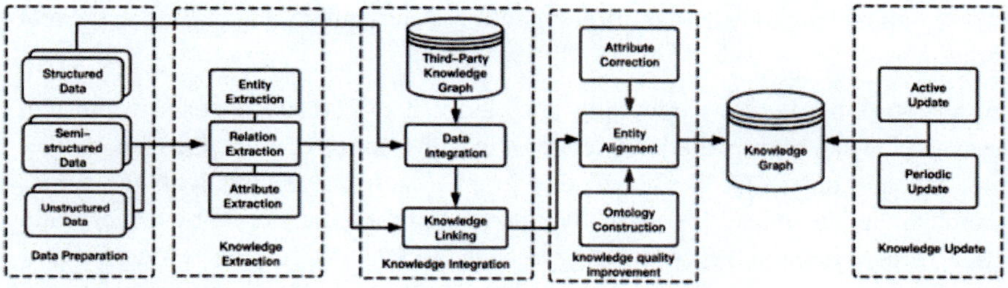

Fig. 1 The layered structure of Mongolian knowledge graph construction

researchers divide it into different stages, the most important stage in the construction of a knowledge graph is as follows: first of all data preparation step, then knowledge extraction step, and knowledge integration step, knowledge quality improvement step, at last knowledge update step. Figure 1 shows a conceptual structure of knowledge graph construction.

Data Preparation First of all, data acquisition is an indispensable stage for Mongolian knowledge graph construction. The data resource mainly can be classified into three types. One is structured data, which means that any data that are stored in a fixed field within a record or file such as data in relational databases and spreadsheets. Semi-structured data means some type of data that have a certain structure but lack the strict data model structure and that lack flexibility, such as HTML, JSON, and XML format data. Unstructured data means the information that cannot be so readily classified, and the data model is not pre-defined, like plaintext.

Knowledge Extraction In the stage of knowledge extraction, it uses some automatic methods to extract information such as named entities, relationships between entities, and attributes of the entity from semi-structured and unstructured data [15]. Entity extraction is also known as named entity recognition, and it aims to automatically identify and classify named entities into pre-defined categories such as person, location, organization, time, and so on from text content, which is the most fundamental and important part of knowledge extraction. The quality of entity extraction is very important that will directly affect the follow-up work. After getting the entities, in order to construct knowledge graphs, it must extract the relationships between the entities from the relevant corpus and analyze the conceptual extract relations. The attribute extraction is finding out the properties of the entities and is to define the intentional semantics of the entities.

Knowledge Integration This step mainly links entities extracted at previous steps between different sources and other equivalent knowledge into the knowledge graph.

Knowledge Quality Improvement The results of previous work may contain erroneous information or noise. Incorrect or erroneous data have to be removed for purpose of ensuring the quality of the knowledge graph. Entity alignment is the process of determining whether different entities refer to the same objects in the real world.

Knowledge Update New emerging knowledge continuously appears in the real world. At the same time, the knowledge of existing entities could also change over time. So the content of the knowledge graph requires evolving with the times. Periodical update means periodically updating the content of knowledge graph with a fresh version and incorporating changes in knowledge after some period of time. The active update refers to frequently updating knowledge graphs with lower cost [16].

2.2 Challenges in Constructing Mongolian Knowledge Graph

The construction of a knowledge graph is a language-dependent problem. Knowledge extraction highly depends on language itself, because the process of information extraction is highly dependent on the lexicon's analysis, syntax, and semantics of all the elements, attributes, and content. There are significant linguistic differences between Mongolian and English or Chinese in various perspectives. Currently, most released projects of knowledge graphs are constructed in English or Chinese. Directly translating these knowledge graphs into Mongolian is not suitable since these languages belong to a different language family.

After investigation, we found the challenges faced by constructing Mongolian knowledge graph are as below:

1. Data sources in Mongolian are limited especially for the domain field. Construction of Mongolian domain-specific knowledge graph needs the data of the certain domain. There are a lot of webpages related to domain fields such as Wikipedia in English or Chinese that are publicly available, but there are no such resources in Mongolian. In the first step of data preparation, it needs great effort, even have to convert printed documents or images into text by OCR tools, and it needs further manual correction.
2. Knowledge extraction algorithms for different languages are not generic, and it is highly dependent on language itself. Mongolian is different from Chinese or English in many aspects such as vocabulary and grammar. New algorithms or tools are needed since it cannot directly apply the existing algorithms or techniques to the Mongolian context.

3 Mongolian Named Entity Recognition

In this section, we introduce a neural-network architecture for Mongolian named entity recognition.

Entity extraction methods are mainly divided into three kinds. First is rule-based approaches, and another one is statistical machine-learning-based approaches. The most commonly used statistical machine-learning-based models are Hidden Markov Model (HMM) [17], Maximum Entropy Model [18], Maximum Support Vector Machine (SVM) [19], and Conditions Random Field (CRF) [20], etc. [21]. The last is the neural-network-based approaches. As in article [22], many manual labeled samples are required in the training process of the model but cannot get the obvious effect.

With the rapid development of machine-learning technology, the techniques that are based on neural-network architectures are attracted much attention from researchers. The article [23] shows that using the Convolutional Neural Networks (CNN) model to deal with the task of feature extraction and get good performance. Huang et al. used a BiLSTM-CRF model for the sequence tagging task and much improved the model performance in 2015 [24]. Santos et al. realized that using word- and character-level embedding DNN achieved good performance for named entity recognition task [25]. Lampleet et al. obtained state-of-the-art performance by using the LSTM-CRF model and S-LSTM that rely on word representations and character-level representations in 2016 [26]. In 2018, Fenget et al. introduced a BILSTM neural network structure-based named entity recognition method [27]. Maimatiayifu et al. performed BiLSTM-CNN-CRF model to NER task in Uyghur language [28]. Li Lishuang et al. introduced a method that implemented the CNN-BiLSTM-CRF model to chemical compound and drug name recognition and got the highest F1 value at that time [29].

Inspired by the works above, we recommend that using the CNN-BiLSTM-CRF model as a Mongolian entity recognition and extraction method. As shown in Fig. 2, the framework of the CNN-BiLSTM-CRF model is composed of three layers. Those are the CNN module, the BiLSTM module, and the CRF module, respectively. The convolution layer in the CNN converts the information of text into a matrix of features in order to extract local information. The most representative part can be obtained as a character vector through the convolution and maximum pooling layer. Existing research shows that Convolutional Neural Network (CNN) is the most efficient technique for morphological feature extraction tasks such as extract and encode word prefix and suffix of word characters as vector representation. The extraction of character-level features is bone by CNN layer in the field of named entity recognition.

Correctness of named entities recognition in a sentence is contingent in the context of a certain word. Before and after the words are significant to predict labels, it is very helpful for named entity recognition task if it can get the past and future of the context information, but LSTM hidden layer only gets information from the past, does not know anything about the future, and the bidirectional LSTM is a better solution. The previous works have demonstrated the effectiveness. The main idea is taking the input sequence in a forward direction and reverse sequence in a backward direction. BiLSTM can get information from past (backward) and future (forward) states simultaneously. The final output is concatenated from the two hidden LSTM layers.

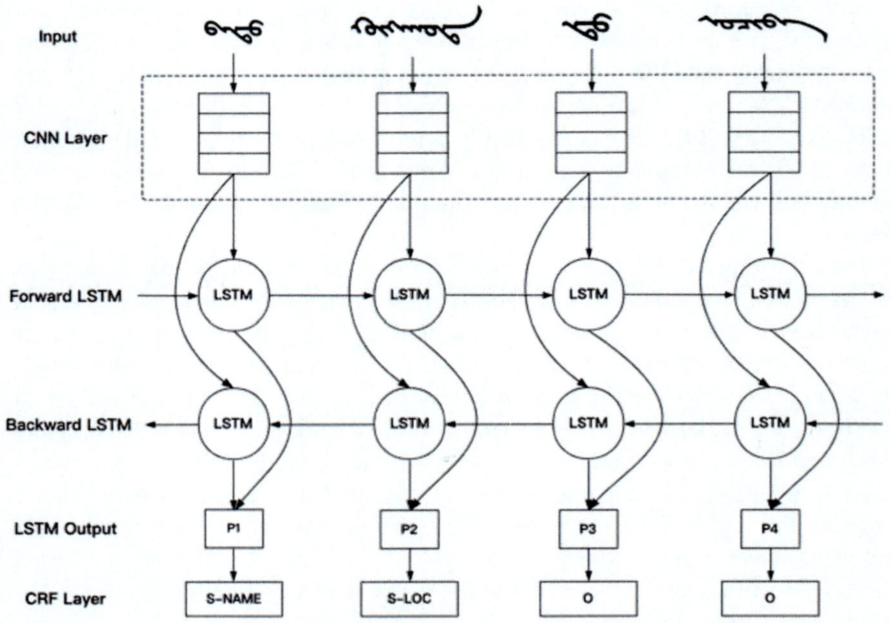

Fig. 2 The CNN-BiLSTM-CRF model of Mongolian named entity recognition. The integral structure of the model consists of three parts: (1) The CNN layer extracted character-level features by using convolution and maximum pooling layer. (2) The BiLSTM layer captures past and future information of the word in the sentence. (3) The CRF layer is labeled with CRF. S in S-NAME denotes the single name, NAME denotes the name of a person, S in S-LOC denotes the single location name, and LOC denotes the name of the location

The last layer of the CNN-BiLSTM-CRF model is the CRF layer. The optimal global annotation sequence is obtained by processing the output results of the BiLSTM module.

For the labeling strategy of supervised learning, the BIO, BIEO, and BMESO, etc., are commonly used. According to the research of [15, 30], the BMESO labeling is better than others because it can more clearly separate the boundary of entities.

For the tag of each entity in the sentence, the first word is marked as "B-entity name," the middle word is marked as "M-entity name," the end word is marked as "E-entity name," the single entity is marked as "S-entity name," and the none entity is marked as "O." The labeling strategy of BMESO is shown in Table 1.

An example of a given Mongolian sentence text entity annotation while using the BMESO labeling strategy is shown in Table 2.

Table 1 Annotation strategy of BMESO

Type	Beginning tag	Middle tag	End tag	Single tag
Time	B-TIME	M-TIME	E-TIME	S-TIME
Name	B-NAME	M-NAME	E-NAME	S-NAME
Location	B-LOC	M-LOC	E-LOC	S-LOC
Organization	B-ORG	M-ORG	E-ORG	S-ORG
...	...	...	...	...
None entity	O	O	O	O

Table 2 A demonstration of Mongolian sentence named entity labeling

English	Batu went to Hohhot			
Text	ᠪᠠᠲᠤ	ᠬᠥᠬᠡᠬᠣᠲᠠ	ᠳᠤ	ᠶᠠᠪᠤᠪᠠ
Tag	S-NAME	S-LOC	O	O

4 Conclusions and Future Work

This chapter provides a survey on constructing a Mongolian vertical domain knowledge graph. First, it briefly introduced the process of constructing knowledge graph. Second, it discussed and analyzed the challenges faced by constructing Mongolian knowledge graph. Last, it recommended a neural-network architectures CNN-BiLSTM-CRF model for Mongolian named entity recognition.

For future work, we will make a corpus using the BMESO annotation strategy for Mongolian resources and then use CNN-BiLSTM-CRF model to extract named entities in unstructured Mongolian plaintext. Moreover, it improved the performance and accuracy of Mongolian named entity recognition. Meanwhile, in order to construct a Mongolian vertical domain knowledge graph, we will focus on the relationship extraction technique between Mongolian entities.

References

1. Singhal, A.: Introducing the knowledge graph: things, not strings. Off. Google Blog **5** (2012). https://www.blog.google/products/search/introducing-knowledge-graph-things-not/
2. Kejriwal, M.: Domain-specific knowledge graph construction. Springer International Publishing, New York (2019)
3. Vrandec ˇić, D., Krötzsch, M.: Wikidata: A free collaborative knowledgebase. Commun. ACM **57**(10), 78–85 (2014). https://doi.org/10.1145/2629489
4. Suchanek, F.M., Kasneci, G., Weikum, G.: YAGO: A core of semantic knowledge. In: Proceedings of the 16th International Conference on World Wide Web, pp. 697–706 (2017). http://www2007.wwwconference.org/papers/paper391.pdf
5. Lehmann, J., Isele, R., Jakob, M., Jentzsch, A., Kontokostas, D., Mendes, P.N., Hellmann, S., Morsey, M., van Kleef, P., Auer, S.: DBpedia—a large-scale, multilingual knowledge base extracted from Wikipedia. Semantic Web **6**(2), 167–195 (2015). https://doi.org/10.3233/SW-140134

6. Bollacker, K., Evans, C., Paritosh, P., Sturge, T., Taylor, J.: Freebase: A collaboratively created graph database for structuring human knowledge. In: Proceedings of the 2008 ACM SIGMOD International Conference on Management of Data, pp. 1247–1250 (2008). https://doi.org/10.1145/1376616.1376746
7. Xu, B., Xu, Y., Liang, J., Xie, C., Liang, B., Cui, W., Xiao, Y.: CN-DBpedia: A never-ending Chinese knowledge extraction system. In: Proceedings of the International Conference on Industrial, Engineering and Other Applications of Applied Intelligent Systems, Arras, France, 27–30 June 2017, pp. 428–438. Springer, Cham (2017)
8. Niu, X., Sun, X., Wang, H., Rong, S., Qi, G., Yu, Y.: Zhishi.me-Weaving Chinese linking open data. In: Proceedings of the Semantic Web–ISWC 2011, Bonn, Germany, 23–27 October 2011, pp. 205–220. Springer, Berlin (2011)
9. MrYener. OwnThink Knowledge Graph. https://www.ownthink.com/ (accessed on 30 March 2020)
10. Wang, Z., Li, J., Wang, Z., Li, S., Li, M., Zhang, D., Shi, Y., Liu, Y., Zhang, P., Tang, J.: XLore: A large-scale english-chinese bilingual knowledge graph. In: Presented at the Meeting of the International Semantic Web Conference (Posters & Demos), Sydney, Australia, 21–25 October 2013 (2013)
11. Lewis, M.P., Simons, G.F., Fennig, C.D.: Ethnologue: Languages of the World, 18th edn. SIL International, Dallas, TX (2015). http://www.ethnologue.com
12. Zhao, Z., Han, S.K., So, I.M.: Architecture of knowledge graph construction techniques. Int. J. Pure Appl. Math. **118**(19), 1869–83 (2018)
13. Yang, S., Han, R.: Method and tool analysis of knowledgemapping abroad. Libr. Inf. Knowl. **6**, 101–109 (2012)
14. Börner, K., Chen, C., Boyack, K.W.: Visualizing knowledge domains. Annu. Rev. Inf. Sci. Technol. **37**, 179–255 (2003)
15. Ratinov, L., Roth, D.: Design challenges and misconceptions in named entity recognition. In: Proceedings of the Thirteenth Conference on Computational Natural Language Learning, Boulder, CO, USA, 4–5 June 2009 (2009)
16. Wu, T., Qi, G., Li, C., Wang, M.: A survey of techniques for constructing Chinese knowledge graphs and their applications. Sustainability **10**(9), 3245 (2018). https://doi.org/10.3390/su10093245
17. Han, X., Huang, D.: Study of Chinese part-of-speech tagging based on semi-supervised hidden markov model. Small Microcomput. Syst. **36**, 2813–2816 (2015)
18. Borthwick, A.E.: A Maximum Entropy Approach to Named Entity Recognition, Ph.D. Thesis. New York University, New York (1999)
19. Wallach, H.M.: Conditional Random Fields: An Introduction. Tech. Rep., vol. 53, pp. 267–272 (2004)
20. He, Y., Luo, C., Hu, B.: A geographic named entity recognition method based on the combination of CRF and rules. Comput. Appl. Softw. **32**, 179–185 (2015)
21. Wang, Z., Jiang, M., Gao, J., Chen, Y.: A Chinese named entity recognition method based on BERT. Comput. Sci. **46**, 138–142 (2019)
22. Wang, Z., Jiang, M., Gao, J., Chen, Y.: A Chinese named entity recognition method based on BERT. Comput. Sci. **46**, 138–142 (2019)
23. Collobert, R., Weston, J., Bottou, L., Karlen, M., Kavukcuoglu, K., Kuksa, P.: Natural language processing (Almost) from scratch. J. Mach. Learn. Res. **12**, 2493–2537 (2011)
24. Huang, Z., Xu, W., Yu, K.: Bidirectional LSTM-CRF Models for Sequence Tagging (2015). arXiv:1508.01991.
25. Santos, C.N, Guimaraes, V.: Boosting named entity recognition with neural character embeddings. arXiv preprint arXiv:1505.05008 (2015)
26. Lample, G., Ballesteros, M., Subramanian, S., Kawakami, K., Dyer, C.: Neural architectures for named entity recognition. arXiv:1603.01360 (2016)
27. Yan-hong, F.E.N.G., Hong, Y.U., Geng, S.U.N., Juan-juan, S.U.N.: Named entity recognition method based on BLSTM. Comput. Sci. **45**(2), 261–268 (2018)

28. Maimai, A., Wushou, S., Palidan, M., Yang, W.: Uyghur named entity recognition based on BILSTM-CNN-CRF model. Comput. Eng. **44**, 230–236 (2018)
29. Li, L.S., Guo, Y.: Biomedical named entity recognition based on CNN-BiLSTM-CRF model. Chin. J. Inf. **32**(1), 116–122 (2018)
30. Dai, H.J., Lai, P.T., Chang, Y.C., Tsai, R.T.H.: Enhancing of chemical compound and drug name recognition using representative tag scheme and fine-grained tokenization. J. Cheminform. **7**(Suppl. 1), S14 (2015)

Mongolian Word Segmentation Based on BiLSTM-CNN-CRF Model

Wuyun He and Siriguleng Wang

1 Introduction

With the maturity of information technology and the development of the Internet, natural language processing for ethnic minorities turns out to be an essential and important research area. Mongolian word segmentation is a primary task in Mongolian natural language information processing and is a significant foundation for subsequent research on Mongolian syntactic analysis, named entity recognition, text retrieval, and machine translation. Treating Mongolian words as a whole will lose a lot of grammatical and semantic information. Segmenting Mongolian words can effectively reduce the loss of grammar and semantics and solve the problem of data sparseness. For Mongolian word segmentation, in 1997, Nashun-Uritu [1] proposed a rule-based approach to achieve an automatic root, stem, and end word segmentation system, which covers more than 95% of all types of modern Mongolian texts. In 2009, Hou Hongxu and Liu Qun et al. [2] proposed a Mongolian word segmentation method based on a combination of rules and statistics, thus achieving an accuracy rate of 93.9%. In 2009, Cong Wei [3] proposed the hidden Markov model (HMM) for Mongolian word segmentation for the features of Mongolian word formation and morphology, thus achieving an accuracy rate of 97.13%. In 2010, Zhao Wei and Hou Hongxu et al. [4] proposed a Mongolian word segmentation system based on a conditional random field model, resulting in an accuracy rate of 99.2%. In 2011, Li Wen et al. [5] used a phrase-based statistical machine translation model for word list words and unregistered words with a word segmentation accuracy of 96.94%. In 2011, Jiang Wenbin et al. [6] proposed a word segmentation slice strategy based on discriminative classification, which has good

W. He (✉) · S. Wang
Inner Mongolia Normal University, Hohhot, Inner Mongolia, China
e-mail: 1136669501@qq.com; siriguleng@imnu.edu.cn

D. Tang et al. (eds.), *Mobile Wireless Middleware, Operating Systems and Applications*, EAI/Springer Innovations in Communication and Computing,
https://doi.org/10.1007/978-3-030-98671-1_9

generalization ability. In 2011, Ming Yu [7] obtained a word segmentation method for Mongolian with excellent performance by combining lexical, rule, and statistical approaches to complement each other; in 2019, Ren Zhong et al. [8] used the byte pair encoding (BPE) algorithm to study the subword granularity slicing technique in Mongolian-Chinese neural machine translation assignments and alleviated the data sparsity problem.

In recent years' research, deep learning has been rapidly developed in the field of natural language processing. In contrast to other machine learning methods, it allows new and effective feature representations to be learned quickly from the training data for new applications. At present, neural network models have now achieved good achievements in word segmentation tasks. In 2011, Collobert et al. [9] proposed a three-layer feedforward neural network modeling to deal with the Chinese word segmentation problem applying neural networks to natural language processing tasks. In 2013, Zheng et al. [10] used deep learning methods to solve the Chinese word segmentation issue. In 2015, Chen et al. [11] used LSTM neural networks to solve the Chinese word segmentation problem and achieved good results. In 2016, Yao et al. [12] used bidirectional LSTM neural networks for Chinese word segmentation. In 2019, Yajing Sun et al. [13] used BiLSTM-CRF neural network model to study word segmentation in Uyghur language and achieved significant improvement. In 2020, Wang Lili et al. [14] used BiLSTM-CRF neural network for Tibetan word segmentation and achieved better results.

The neural network model has achieved good results in the application of word segmentation tasks in various languages, but the Mongolian word segmentation tasks application is relatively small compared with other languages. In 2019, Chen Shengai [15] proposed a deep neural network-based Mongolian word segmentation method and constructed a framework based on the BiLSTM-CRF model, and the word segmentation accuracy reached 97.08%. Suyala et al. [16] used transformer-CRF algorithm to segment Mongolian words and applied it to the Mongolian-Chinese machine translating, and the translation performance has been improved.

Mongolian is a cohesive language with many morphological changes such as number, case, time, aspect, and state. Mongolian words are formed by adding various additional elements to the stem. The additional elements of Mongolian are composed of additional elements of word formation and additional elements of configuration. The additional component of word formation will change the meaning of the original stem, but the additional component of formation will not change the meaning of the stem. The additional component of configuration is the morphological change of Mongolian words. It does not modify the definition of the word but only increases the grammatical meaning. Therefore, in view of the Mongolian adhesiveness and the rich and complex characteristics of word shape changes, this paper combines the BiLSTM-CRF model with the CNN model to optimize, builds a multilayer neural network, and conducts a segmentation study on the additional components of the Mongolian structure. Through comparative experiments, it is demonstrated that the deep neural network model is superior to Mongolian word segmentation.

2 Neural Network Model

2.1 BiLSTM Model Layer

The LSTM model is a particular form of recurrent neural network, and the structure of the LSTM model is shown in Fig. 1. Mainly through training, it is possible to understand which information needs to be remembered and which information needs to be forgotten, which can effectively model distant dependencies. However, as it only remembers information from the past, it cannot consider future contextual information. To effectively address this problem, the BiLSTM model [17], consisting of a forward LSTM and a backward LSTM, is proposed, which is capable of capturing bidirectional Mongolian character feature information. In this model framework, local semantic feature vectors are spliced behind the current character vectors to obtain a set of fusion vectors, which are then fed into the BiLSTM so that the model learns the contextual features of the character sequences. In addition, the dropout layer is added to the input and output of the BiLSTM model, thus preventing overfitting of the model and improving the overall performance of the model. The BiLSTM model layer is connected to the softmax linear layer to automatically extract the features of Mongolian sentences and finally acquire the corresponding label information for each Mongolian character.

The LSTM cell is updated at moment t by the following equation:

$$f_t = \sigma\left(W_f \cdot [h_{t-1}, x_t] + b_f\right) \tag{1}$$

$$i_t = \sigma\left(W_i \cdot [h_{t-1}, x_t] + b_i\right) \tag{2}$$

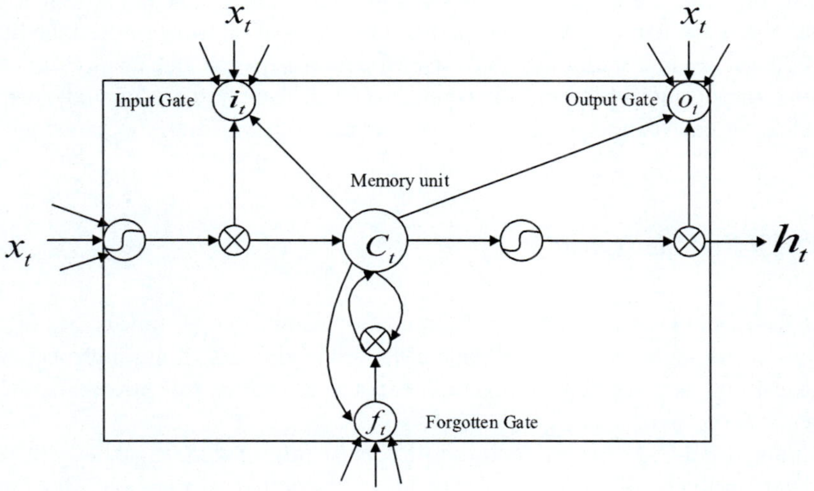

Fig. 1 LSTM model structure diagram

$$\tilde{C}_t = \tanh\left(W_c \cdot [h_{t-1}, x_t] + b_c\right) \tag{3}$$

$$C_t = f_t \times C_{t-1} + i_t \times \tilde{C}_t \tag{4}$$

$$o_t = \sigma\left(W_o\left[h_{t-1}, x_t\right] + b_o\right) \tag{5}$$

$$h_t = o_t \times \tanh\left(C_t\right) \tag{6}$$

where i_t is the input gate, f_t is the forgetting gate, C_t is the memory unit, o_t is the output gate, h_t and x_t are the hidden layer vector and the input vector at time t, σ is the sigmoid activation function, tanh is the hyperbolic tangent activation function, and W and b are the weight matrix and bias vector, respectively.

Although the BiLSTM is able to learn contextual information, the outputs are independent of each other. Softmax classifiers only select the label outputs with the highest probability, resulting in words with incorrect grammatical structure being produced. Therefore, to obtain the globally optimum sequence, a CRF network layer is added after the BiLSTM layer.

2.2 CNN Model Layer

CNN models generally consist of an input layer, several convolutional layers, several pooling layers, a fully connected layer, and an output layer [18]. Research has shown that CNNs are very effective in extracting local features [19], and the use of CNN networks can be effective in reducing model parameters and training time. In this model framework, the local feature extraction model of CNN is shown in Fig. 2. The convolutional layer mainly performs convolutional operations, the pooling layer uses the feature of repetition of local information of the data to represent the output of the whole region with overall features for local information, and the fully connected layer is a traditional feedforward neural network that mainly expands multiple outputs into a layer of fully connected form. Therefore, in this study, fusion models are used to extract local feature representations of words using CNN models.

2.3 CRF Model Layer

The CRF model is an undirected graph model that combines the advantages of the maximum entropy model and the hidden Markov model [20]. It has high accuracy in the tasks of word segmentation, lexical annotation, and named entity recognition sequence annotation.

Although BiLSTM makes good use of context information, it cannot take full advantage of the predicted label information of the output information. Therefore, the CRF layer is added after the BiLSTM model layer, the output information of BiLSTM is used as the input information of the CRF layer, the entire sequence is

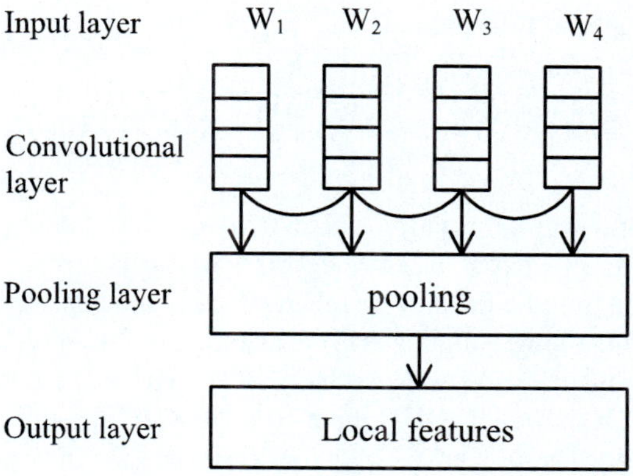

Fig. 2 Structure of CNN model for extracting local features

scored through CRF, and some constraints are added to ensure the rationality of the predicted label. For example, the label of the first character in a word must start with the label "B" or "S." Therefore, the experiment in this paper uses the CRF model to perform joint decoding, instead of decoding each tag individually.

3 BiLSTM-CNN-CRF-Based Mongolian Word Segmentation Model

In the Mongolian-Chinese machine translation task, the segmentation of the corpus is a very important preprocessing step. Mongolian word segmentation belongs to the character-level classification problem in the sequence labeling task, which is to classify each character. We optimize a variety of neural network models by fusion, so that the model can discriminate and classify each character in the data, so as to achieve the effect of word segmentation. The following first introduces the training set form and the training data set labeling method used in this research and then gives the overall framework of the model.

3.1 Data Annotation

The commonly used Mongolian script is divided into traditional Mongolian script and Latin Mongolian script. These two Mongolian scripts present two completely different written forms. The correct conversion between traditional Mongolian and Latin Mongolian is an essential step in data labeling. For the convenience of processing, the experiment converts all the traditional Mongolian data into Inner Mongolian form with the conversion tool developed by our laboratory [21].

In word segmentation, each character in the training corpus needs to be single-point marked. In this experiment, according to Mongolian language characteristics

Table 1 Examples of labels

Character sequence	H	O	G	J	I	G	U	L	H	U
Label sequence	B	M	M	M	E	B	M	E	B	E

and word formation characteristics, the character categories are divided into {B, M, E, S} four label categories, where B represents the root (head), M represents the middle of the word, E represents the configuration suffix (suffix), and S stands for a single character. For example, a Mongolian word is "ᠬᠣᠭᠵᠢᠭᠤᠯᠬᠤ," and the Latin transliteration form is "HOGJIGULHU." The corresponding segmentation result is "HOGJI/GUL/HU," the root is HOGJI, and the configuration suffix are GUL and HU. Using {B, M, E, S} four types of labeling results as shown in Table 1, the first line is a single Mongolian character sequence converted from Mongolian words, and the second line is the labeling sequence corresponding to Mongolian characters.

3.2 *Character Vectorization*

Before the model training, the Mongolian characters are vectorized. This paper adopts the word2vec model open sourced by Google in 2013. Word2vec maps each Mongolian character in a sentence into a dense low-dimensional character vector. Word segmentation method based on BiLSTM-CNN-CRF considers the information before and after the character in the training process, mapping the latter vector can express the semantic distance by computing the distance between the vectors. In the process of character vectorization, during character vectorization, we set the standard length of the input sequence to 30, and the sequence is automatically filled to zero if it is not long enough.

3.3 *BiLSTM-CNN-CRF Model Layer*

In the paper, based on the BiLSTM-CRF, a commonly used sequence labeling model in the Chinese and English domains, to increase sequence labeling accuracy, a CNN network structure is added. The BiLSTM model is applied to extract the before and after entering the information for the character sequence; the CNN model is used to extract the local feature information of characters; the CRF model is used for the final step of sequence labeling. That is, the BiLSTM-CNN-CRF deep learning model is constructed for Mongolian word segmentation task, and the structure diagram is shown in Fig. 3.

Figure 3 shows that the output vector of BiLSTM is input to the CRF layer to construct and create a neural network model. The neural network architecture of this paper is consisted of BiLSTM module, CNN module, and CRF module. The

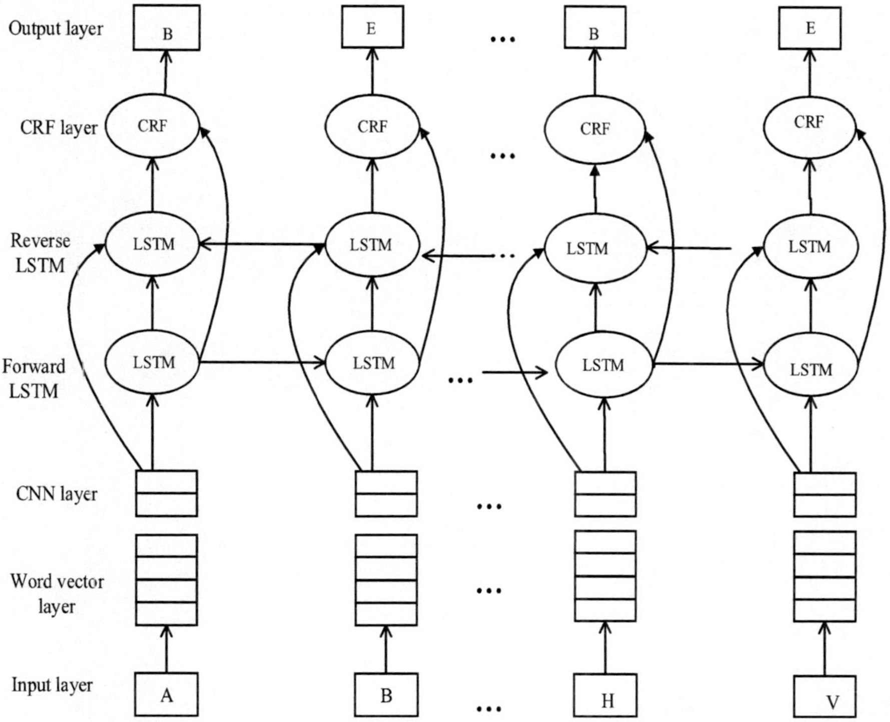

Fig. 3 BiLSTM-CNN-CRF model framework

input to the model is "Mongolian characters," and the output is "sequence labels." The character vector is first pre-trained by using the input layer data. In order to facilitate the post-processing, the current Mongolian sentence is converted into Latin Mongolian by the conversion algorithm. The CNN module performs convolution and maximum pooling on the character vector matrix to obtain the local feature representation of each word. The mixed vector of the character-level local feature representation of each word and the pre-trained character vector is used as the input of the next layer of neural network model BiLSTM. Finally, the output of the BiLSTM was added as input to the CRF network layer, and the optimal annotation sequence was jointly decoded and output.

4 Experiments and Analysis

In the paper, four neural network models, LSTM, BiLSTM, BiLSTM-CRF, and BiLSTM-CNN-CRF, are applied to the Mongolian word segmentation task to prove each model's performance, and the Mongolian word segmentation application model is constructed as shown in Fig. 4.

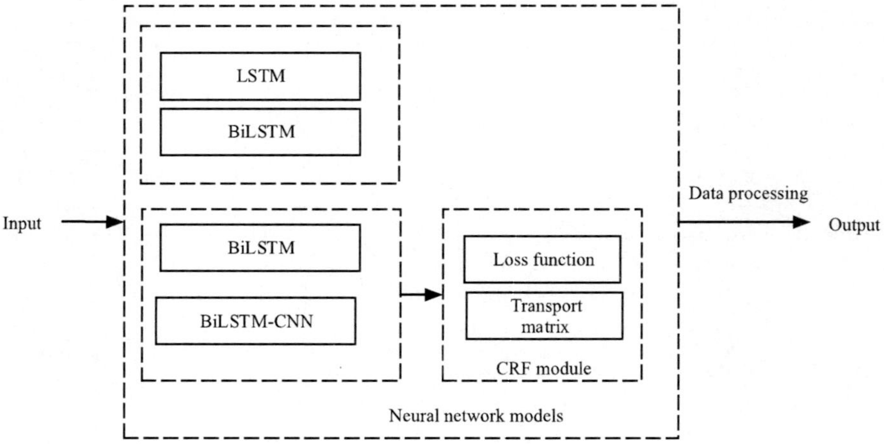

Fig. 4 Word segmentation model

Table 2 Mongolian word segmentation data set details

Data set	Number of lines	Number of words
Training set	45,017	769,413
Test set	5000	82,352

4.1 Experimental Data

The word segmentation experiment uses the ethnic minority language word segmentation technology to evaluate the Mongolian training evaluation corpus provided by the MLWS2017 (Minority Language Word Segmentation) activity. The scale is 50,017 sentences. The data set is randomly separated into training set and test set. The ratio of training set to test set is 9:1, and the maximum sentence length in the training corpus is limited to 100, so as to avoid the model spending too much training time on long sentences. The data details are shown in the following Table 2.

4.2 Experimental Parameter Settings and Evaluating Indicator

To ensure optimum fit, it is essential to use an optimization function to optimize the weights and biases between the neural units of each layer, so that the quantized training input is as close as possible to the target output of the model and the optimal word segmentation result is achieved. The TensorFlow framework is selected based on the BiLSTM-CNN-CRF training model, and the specific parameter configuration is: batchSize is set to 64; maxlen is set to 30, the number of hidden layer units of forward LSTM and backward LSTM is set to 100, and the character embedding The dimension is set to 100, the learning rate is set to 0.01, and the dropout size is set to

0.2, which means that 20% of the neurons are randomly dropped when training the model.

The evaluation metrics for Mongolian word segmentation were evaluated using a standard evaluation method, namely, the three combined metrics of precision (P), recall (R), and F value. The calculation formulae are shown below.

The precision rate is determined by the percentage of the correct number of word segmentation to the total number of word segmentation. In the word segmentation task, we focus on how many correct segmentation results we want to know, so this article uses the precision as evaluation indicator.

$$\text{Precision} = \frac{\text{Number of correctly segmented lexemes}}{\text{Total number of lexemes segmented}} \times 100\% \tag{7}$$

The recall rate is different from the precision rate. It is determined by the percentage of correct word segmentation to the total number of word segmentation in the test set. It focuses on describing how many of the correct segmentation results are successfully selected, and it is also one of the important indicators for evaluating word segmentation tasks.

$$\text{Recall} = \frac{\text{Number of correctly segmented lexemes}}{\text{Total number of lexemes that should be segmented}} \times 100\% \tag{8}$$

The precision rate and recall rate are usually high and the other low, which means that if one of the indicators is high, the other is low. The F value is the result of a weighted average of accuracy and recall, which can be seen as a combination of the two and can represent the overall quality of the word segmentation of the whole model system. In general, a higher *F* value indicates a better methodology and effectiveness of the experiment.

$$F\ \text{value} = \frac{2 \times \text{P} \times \text{R}}{\text{P} + \text{R}} \times 100\% \tag{9}$$

4.3 Experimental Results and Analysis

The experiments are carried out in three main aspects: (1) using character vectors of different dimensions in the BiLSTM-CNN-CRF model to obtain the character vector dimension that is most applicable to the model; (2) using different dropout values in the BiLSTM-CNN-CRF model to make the model optimal; and (3) comparing the performance of different neural network models on the Mongolian word sorting task and evaluating the effectiveness of the BiLSTM-CNN-CRF model based on the BiLSTM proposed in this paper for the Mongolian word segmentation task.

Table 3 BiLSTM-CNN-CRF model with different character vector dimensions

Dimensionality	*P*/%	*R*/%	*F*/%
50	98.79	98.67	98.73
100	99.47	99.46	99.47
150	99.04	99.08	99.06
200	98.45	98.53	98.49

Table 4 BiLSTM-CNN-CRF models with different dropout ratios

Dropout	*P*/%	*R*/%	*F*/%
20%	99.47	99.46	99.47
30%	99.08	99.14	99.11
50%	98.74	98.76	98.75
80%	97.60	97.54	97.57

1. Impact of Character Vector Dimensionality on Model Performance

The dimensionality of the character vectors is crucial to the accuracy of Mongolian word segmentation and the speed of model training. In the experiments, character vectors of 50, 100, 150, and 200 dimensions were chosen for word segmentation in Mongolian while keeping the other parameters of the BiLSTM-CNN-CRF model unchanged, and the results were evaluated using accuracy, recall, and F value, as shown in Table 3.

Table 3 shows that the performance of the model with a character vector dimension of 100 is relatively better than that of the model with dimensions of 50 and 150, with higher *P*, *R*, and *F1* values than the other two dimensions. When the character vector dimension was set to 50, it could not completely cover the word features, and when the character vector dimension was set to 150, the training time of the model became longer and the performance decreased. Explain that in Mongolian word segmentation, the character vector dimension should not be too small or too large.

2. Effect of Regularization on Model Performance

The idea of regularization is to avoid the problem of overfitting by simplifying the model, and dropout is one of the effective ways to do this. Dropout values indicate that they are temporarily removed from the deep neural network during training with a certain probability, thus preventing overfitting of the deep neural network. In this paper, experiments were conducted with the same size corpus at 20%, 30%, 50%, and 80% dropout value ratios with a word vector dimension of 100. The various dropout ratio performance indicators are shown in Table 4.

From the experimental results in Table 4, it can be seen that when the proportion of dropout value is set to 20%, the performance of the model is the best, and the F value reaches 99.47%; when the proportion of dropout value is set to 80%, the F value is significantly reduced, reaching 97.57%. From the experimental results, it can be seen that dropout can immediately delete some nodes without overfitting or underfitting.

Table 5 Results of Mongolian word slicing with different neural network models

Models	*P*/%	*R*/%	*F*/%
LSTM	72.82	73.76	73.29
BiLSTM	97.45	97.39	97.42
BiLSTM-CRF	98.77	98.63	98.70
BiLSTM-CNN-CRF	99.47	99.46	99.47

3. The Effect of Different Neural Network Models on Word Slicing

This paper uses LSTM, BiLSTM, BiLSTM-CRF, BiLSTM-CNN-CRF, and other models for Mongolian word segmentation experiments. First, take the LSTM model as the baseline experiment and compare the experimental analysis with the three models of BiLSTM, BiLSTM-CRF, and BiLSTM-CNN-CRF. To compare the performance of different models on the Mongolian word segmentation task, the same data set is used for Mongolian word segmentation experiments. The performance of the four models is displayed in Table 5.

Table 5 shows that BiLSTM model is more suitable for Mongolian word segmentation tasks than the LSTM model, because the BiLSTM model makes good use of the context. Adding the CRF layer to the model structure can not only learn the context information of sequence tags but also learn the constraint rules between tags, thereby improving the word segmentation effect of the model. On this basis, CNN is introduced as the local feature extractor of Mongolian words, so that the model can be effective in learning the local features of Mongolian words, thereby improving the segmentation effect of Mongolian words. On the task of Mongolian word segmentation, compared with other models, this model has a certain improvement in word segmentation effect, and the *F* value reaches 99.47%.

In the Mongolian word segmentation task, a CRF layer is added to the model structure to learn not only the contextual information of the sequence tags but also the constraint rules between the tags, thus improving the word segmentation effect of the model. On this basis, CNN is introduced as a local feature extractor for Mongolian words so that the model can find local features of Mongolian words effectively and thus improve the cutoff effect of Mongolian words.

5 Conclusion

This research constructs a Mongolian word segmentation framework by fusing BiLSTM model, CNN model, and CRF model. The feature information of the words before and after the current word in the learning layer is obtained through the BiLSTM model, add the CRF layer to the model structure, learn the context information of the sequence label and the constraint rules between the labels, and, on this basis, introduce the CNN model as the local feature of the sequence extractor. Experiments have verified that when the word vector dimension is 100 and the dropout value is 20%, the neural network model based on BiLSTM-CNN-CRF has the best performance on the test data set.

The main role of correct word segmentation in minority languages is also reflected in machine translation, text retrieval, etc., in order to achieve a higher degree of understanding and use of language and even culture. In the future, on the basis of this research, further research and realization of Mongolian word segmentation, part-of-speech tagging, and named entity recognition integration of Mongolian sequence tagging framework will provide services for subsequent Mongolian natural language processing research.

Acknowledgments The research was supported by a grant from the National Natural Science Foundation of China (61762072); Inner Mongolia Autonomous Region Science and Technology Program Project (2021GG0139); and Inner Mongolia Normal University Graduate Research and Innovation Fund Project (GXJJS20129).

References

1. Nashun-Uritu: An automatic cutting system for Mongolian roots, stems and endings. J. Inner Mongolia Univ. (Humanit. Soc. Sci. Ed.). **02**, 53–57 (1997)
2. Hou Hongxu, Liu Qun, Nashun-Uritu, et al.: Mongolian word segmentation based on statistical language models. Pattern Recognit. Artif. Intell. **22**(01), 108–112 (2009)
3. Cong Wei: A Study on Mongolian Word Syncopation System Based on Cascading Hidden Markov Models. Inner Mongolia University, Hohhot (2009)
4. Zhao Wei, Hou Hong Xu, Cong Wei, et al.: A study of Mongolian word segmentation based on conditional random fields. J. Chin. Inf. **24**(05), 31–35+84 (2010)
5. Li Wen, Li Miao, Liang Qing, et al.: Phrase-based statistical machine translation model for Mongolian morphological segmentation. J. Chin. Inf. **25**(04), 122–128 (2011)
6. Jiang Wenbin, W., Jinxing, U.L., et al.: Discriminative stemming of Mongolian with a directed graph morphological analyzer. Chin. J. Inf. **25**(04), 30–34 (2011)
7. Ming Yu: A Study of Mongolian Word Syncopation System Based on Lexicon, Rules and Statistics. Inner Mongolia University, Hohhot (2011)
8. Ren Zhong, Hou Hong Xu, Jia Tu, et al.: Application of subword granularity slicing in Mongolian-Chinese neural machine translation. J. Chin. Inf. **33**(01), 85–92 (2019)
9. Collobert, R., Weston, J., Bottou, L., et al.: Natural language processing (almost) from scratch. J. Mach. Learn. Res. **12**(Aug), 2493–2537 (2011)
10. Zheng, X.Q., Chen, H.Y., Xu, T.Y.: Deep learning for Chinese word segmentation and POS tagging. In: Proceedings of the 2013 Conference on Empirical Methods in Natural Language Processing, pp. 647–657. Association for Computational Linguistics, Seattle (2013)
11. Chen, X.C., Qiu, X.P., Zhu, C.X., et al.: Long short-term memory neural networks for Chinese word segmentation. In: Proceedings of the 2015 Conference on Empirical Methods in Natural Language Processing, pp. 1385–1394. Association for Computational Linguistics, Lisbon (2015)
12. Yao, Y., Huang, Z.: Bi-directional LSTM recurrent neural network for Chinese word segmentation. In: International Conference on Neural Information Processing, pp. 345–353. Springer, Cham (2016)
13. Sun Yajing, Li Chenghua, Yang Bin, et al.: A study on Uyghur word separation based on BI-LSTM-CRF model. J. Qinghai Normal Univ. (Nat. Sci. Ed.). **35**(04), 5–12 (2019)
14. Wang Lili, Wang Hongyuan, Baima Quzhen, et al.: A word separation method for Tibetan based on BiLSTM_CRF model. J. Chongqing Univ. Posts Telecommun. (Nat. Sci. Ed.). **32**(04), 648–654 (2020)

15. Chen Shengai: A Deep Neural Network-Based Analysis of Mongolian Morphological Elements. Beijing Jiaotong University, Beijing (2019)
16. Suila, Z.Z., Renqing Daoji, et al.: Transformer-CRF word slicing method in Mongolian-Chinese machine translation. J. Chin. Inf. **33**(10), 38–46 (2019)
17. Mikolov, T., Sutskever, I., Chen, K., Corrado, G., Dean, J.: Distributed representations of words and phrases and their compositionality. In: Proceedings of NIPS (2013)
18. Hu Xiaohui, Zhu Zhixiang: Research on Chinese word separation methods based on deep learning. Comput. Digital Eng. **48**(3), 627–632 (2020). https://doi.org/10.3969/j.issn.1672-9722.2020.03.025
19. Chiu, J.P.C., Nichols, E.: Named entity recognition with bidirectional LSTM-CNNs. arXiv. **4**, 357–370 (2016)
20. Lafferty, J., Mccallum, A., Pereira, F.: Conditional random fields: probabilistic models for segmenting and labeling sequence data. In: Proc. 18th International Conf. On Machine Learning (2001)
21. Lai, S., Liu, K., He, S., et al.: How to generate a good word embedding. IEEE Intell. Syst. **31**(6), 5–14 (2016)

Safety Helmet Wearing Recognition Based on YOLOv5

Yuhang Ma and Yinfeng Fang

1 Introduction

There are many hidden dangers in the process of construction operations, which makes the accident rate remains high. In recent years, in order to ensure the safe operation of equipment, increasing monitoring systems have been developed based on computer vision. This measure can not only reduce the workload of manual monitoring but also highlight the dangerous operations and avoid the occurrence of accidents [1].

Practice shows that monitoring the behavioral ability and safety equipment of construction workers before the operation, especially the detection of whether helmets are worn correctly, can effectively reduce the accident damage and the occurrence of accidents. However, the current traditional monitoring and safety equipment wearing detection largely rely on the observation and inspection by the experienced managers on site, which generally exists in the phenomenon of low automation level, large workload, and limited inspection items.

With the background of Internet of things and big data, smart construction site can realize the integration of artificial intelligence such as safety helmet and safety rope and add the functions of automatic detection, automatic alarm, and real-time collecting information. It also establishes a big data management platform for the information it transmits and then makes use of the data for security analysis and comprehensive control [2]. Consequently, it's the key for deployment to selecting a suitable object detection algorithm. After years of development, helmet wearing detection has shifted from traditional machine learning method to deep learning method. With the continuous research of deep learning, object detection based on

Y. Ma (✉) · Y. Fang
School of Communication Engineering, Hangzhou Dianzi University, Hangzhou, China

D. Tang et al. (eds.), *Mobile Wireless Middleware, Operating Systems and Applications*, EAI/Springer Innovations in Communication and Computing,
https://doi.org/10.1007/978-3-030-98671-1_10

algorithms such as CNN, SSD, and YOLO has been widely used [3–6]. Among them, the YOLO algorithm is constantly updated; it has a good prospect in the field of object detection because of its real time and accuracy.

Jie Li et al. proposed a practical safety helmet wearing detection method based on machine learning. Firstly, the ViBe background modeling algorithm is exploited to detect the moving object. After obtaining the motion area of interest, the features of histogram of oriented gradient (HOG) are extracted to describe inner human, and the support vector machine (SVM) is trained to classify pedestrians. Finally, the safety helmet detection will be realized by color feature recognition [1]. Mneymneh et al. proposed a framework to monitor helmet wearing. The standard deviation matrix (SDM) is used to detect moving objects, and then the target detector based on aggregate channel features is used to classify people [7]. Jixiu Wu et al. used SSD algorithm to detect safety helmet wearing and developed a novel aggregation framework combined with the presented reverse progressive attention (RPA), which gradually propagates semantically strong features back to the bottom layer. It effectively improves the detection efficiency of small targets [8].

Although current-based deep learning helmet detection methods work well, they still suffer from the following problems: (1) poor performance in complex environments with changeable weather and obstructions. (2) Safety helmets are small targets in complex scene; it is difficult to detect all of them. (3) Different colors of helmets can represent different types of work; previous methods are difficult to classify colors based on the detection of helmets. In view of above problems, this paper adopts the YOLOv5 detection algorithm, which is the latest achievement of YOLO series. On the basis of YOLOv3 [9], YOLOv5 integrates and innovates various advanced algorithms and adds the focus structure to process input images. CSPNet [10] and FPN + PAN structure constitutes the main detection network structure. In this paper, the prediction layer of shallow network is added on the basis of the original network. By training on the dataset, the difficult problem of safety helmet detection can be solved effectively.

2 Method

2.1 Common Target Detection Models

Faster R-CNN

Faster R-CNN is a two-stage object detection algorithm. This algorithm proposes an RPN candidate box generation algorithm, which greatly improves the speed of object detection. Faster R-CNN consists of two modules. The first module is a deep fully convolutional network that proposed regions, and the second module is the Fast R-CNN detector [12]. Faster R-CNN network takes as input an entire image and a set of object proposals. Firstly, the whole image is processed by several convolutional (conv) and max pooling layer to generate a conv feature map. Then,

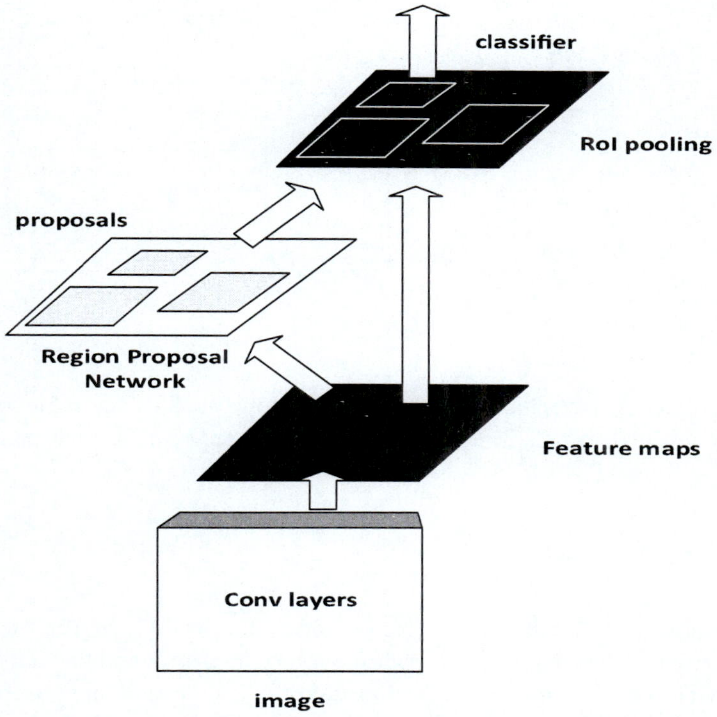

Fig. 1 Faster R-CNN model structure [11]

RPN (region proposal network) processes the extracted convolutional feature maps. RPN is used to look for regions that may contain a predefined number of objects. After getting the possible related objects and their corresponding positions in the original image, the features extracted by CNN and the bounding box containing the relevant objects are used for ROI pooling, and the features of the relevant objects are extracted to obtain a new vector. Finally the network outputs the result, classifies the content in the bounding box, and adjusts the bounding box coordinates (Fig. 1).

SSD

SSD is a one-stage object detection algorithm, which was proposed by Liu et al. at the 2016 ECCV conference [13]. It has a faster detection rate than the two-stage method. The SSD method is based on a feedforward convolutional network, which generates a fixed-size bounding box set and the corresponding score of the target category in the box and then generates the final detection result according to the non-maximization suppression step. The backbone network of SSD is composed of the VGG-16 convolutional network, which adds a part of the convolutional layer on the basis of VGG-16 network to obtain feature maps of different scales for location and

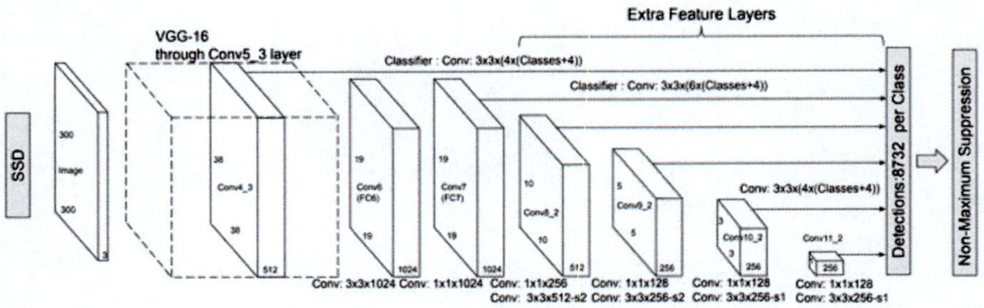

Fig. 2 SSD network structure [13]

category prediction and achieve multi-scale prediction. The SSD algorithm network is shown in Fig. 2, in which the feature map sizes extracted by multi-scale prediction are 38×38, 19×19, 10×10, 5×5, 3×3, and 1×1.

YOLO

YOLO is also a one-stage object detection algorithm. YOLO has been released in five versions, of which YOLOv1 laid the foundation for the entire series, and the following series are improvements based on the first version, only to improve performance. YOLO is short for You Only Look Once [14]. The core idea is to use the whole image as the input of the network through the CNN network and directly regress the position and category of the bounding box. The following will focus on the network structure and the key technologies of YOLOv5.

2.2 YOLOv5

YOLOv5 Network Structure

The structure of YOLOv5 network consists of four parts: input, backbone module, head module, and detect module.

1. Input: Image preprocessing, using mosaic data enhancement, adaptive picture scaling technology, and adaptive anchor box calculation. Mosaic data enhancement is the core of dataset preprocessing. Mosaic data enhancement combines multiple images in random zoom, random crop, and random arrangement, which enriches the dataset and reduces GPU memory (Fig. 3).
2. Backbone: A convolutional neural network that aggregates and forms image features at different image granularities. After the input is sent to the image, the focus slice operation is adopted to make the feature extraction more sufficient.

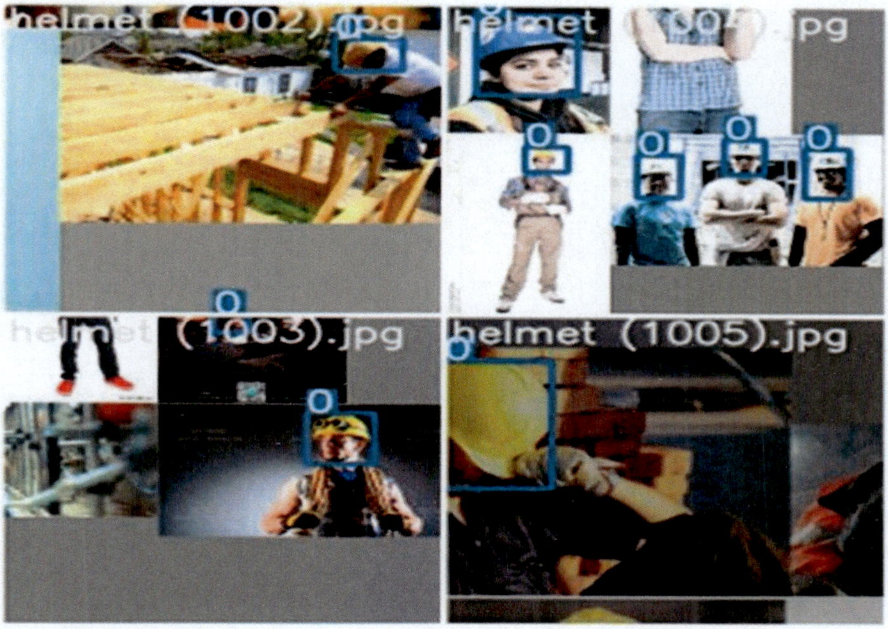

Fig. 3 Mosaic data enhancement

Meanwhile, CSPNet structure is adopted, which has a strong feature extraction ability.

3. Head: A series of network layers that mix and combine image features and then transmit the image features to the prediction layer. Head adopts FPN + PAN network structure. The FPN layer conveys strong semantic features from the top to the bottom, while the feature pyramid conveys strong positioning features from the bottom to the top. They work together; the features of different detection layers are aggregated from different backbone layers.
4. Detect: Detect module predicts image features, generates bounding boxes, and predicts categories. The original network of YOLOv5 outputs three feature maps of different scales, and the prediction terminal uses multiple scales to detect targets of different sales (Fig. 4).

Under the condition of maintaining the same network components, YOLOv5 is divided into four network models with different depths, namely, YOLOv5s, YOLOv5m, YOLOv5l, and YOLOv5x. YOLOv5s is the smallest model, and its average precision is the lowest, but it has the fastest detection speed. On this basis, the other three networks continue to deepen and widen the network, and average precision is continuously improved, but the inference speed is gradually slowing down. Considering the size of the model and the application occasions, this paper uses the YOLOv5s network model for testing (Table 1).

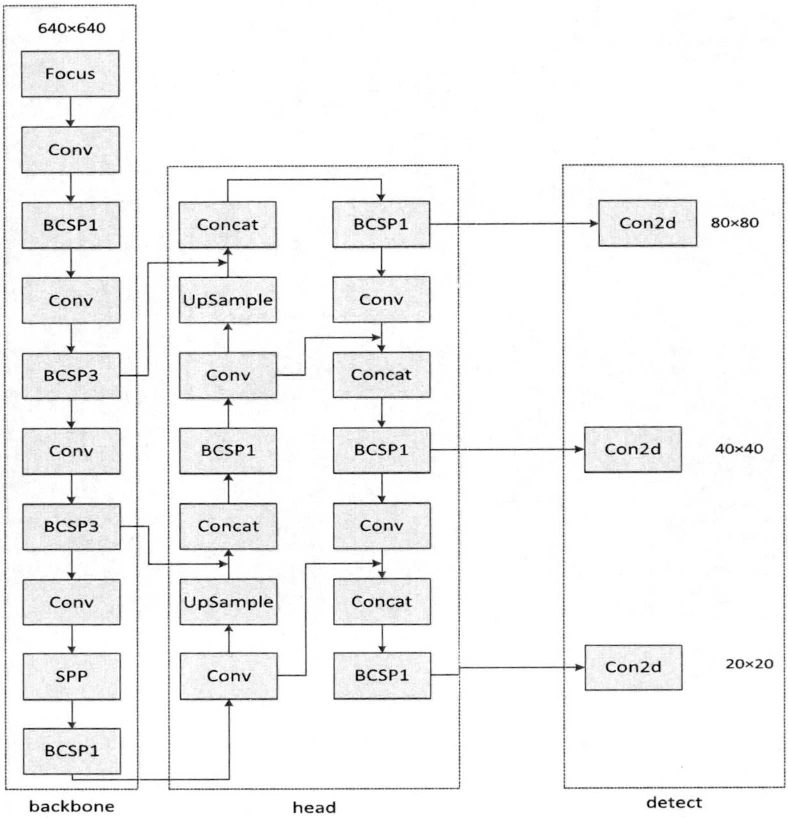

Fig. 4 YOLOv5 network structure

Table 1 Four different depths of YOLOv5 network

	YOLOv5s	YOLOv5m	YOLOv5l	YOLOv5x
depth_multiple	0.33	0.67	1.0	1.33
width_multiple	0.50	0.75	1.0	1.25

The depth_multiple and width_multiple parameters are used to control the depth and width of different network structures. Depth_multiple controls the number of BottleneckCSP in the network, and width_multiple controls the number of convolution kernels in the network.

Enhanced YOLOv5 Network Model

Due to the problems of shooting distance and angle in many images in the experimental dataset, a large number of the helmets appear in the images as small targets.

In an image, there may be a relatively dense number of people wearing helmets. Therefore, in order to enhance the detection performance of small targets such as helmets, this paper proposes an enhanced YOLOv5 network structure. YOLOv5 uses the CSPNet deep neural network for training and detection. It identifies targets of different sizes on three feature maps. However, feature information of some small-sized targets may be lost during detection, resulting in missed detection, false detection, or repeated detection. Generally speaking, the features in the shallow layer are mostly complete, and the outline of the target can be clearly observed, while the features in the deep layer are smaller and more detailed. On the basis of the original YOLOv5 network structure, if the features extracted from the shallow layer network were added into the category judgment and border prediction, a larger and more complete feature map can be obtained, which is beneficial to the detection of small targets.

YOLOv5 detects objects of different sizes at three prediction scales of 20 × 20, 40 × 40, and 80 × 80. Under the condition of keeping the backbone network unchanged, a prediction scale of 160 × 160 was added in front of the third prediction scale 80 × 80 of the original YOLOv5. This improvement combined the strong semantic features of deep layer network with the high resolution of shallow layer network, the original three prediction channels became four prediction channels, and the enhanced YOLOv5 network was built to achieve multi-scale detection on 20 × 20, 40 × 40, 80 × 80, and 160 × 160. It strengthens the detection effect of small targets in the image and improves the detection performance of YOLOv5. The modified network structure is shown in the Fig. 5.

YOLOv5 Loss Function

The YOLOv5 loss function includes classification loss, localization loss (forecast the error between the bounding box and the ground truth), and confidence loss. YOLOv5 used the binary cross entropy loss function to calculate the loss of category probability and target confidence score. This paper uses the GIOU loss function to measure the effect of model convergence:

$$\text{IOU} = \frac{|A \cap B|}{|A \cup B|} \tag{1}$$

$$\text{GIOU} = \text{IOU} - \frac{|C \backslash (A \cup B)|}{|C|} \tag{2}$$

where A is the prediction box, B is the ground truth, and C is the minimum bounding box of A and B.

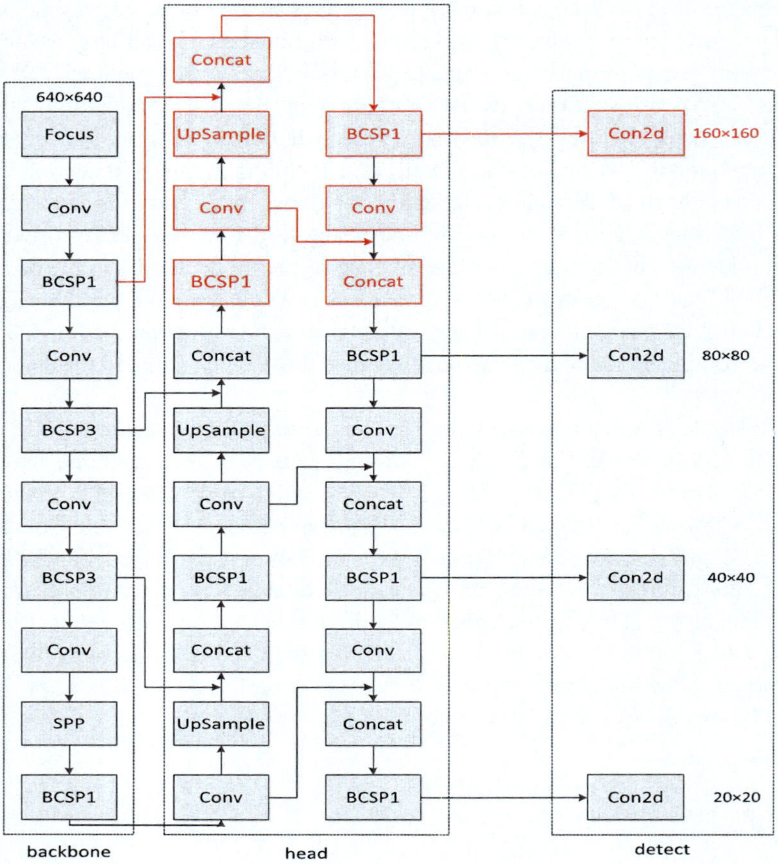

Fig. 5 Enhanced YOLOv5 network structure

3 Experiment

3.1 Training Environment

This experiment builds a test environment under Windows10 system, i7 processor, GTX 960 M GPU with 4G memory. It is based on the PyTorch deep learning framework and implemented through Python3.8 programming. The training process uses GPU processors.

3.2 Dataset

In order to verify the effectiveness and feasibility of the algorithm, the experiments use two different helmet datasets.

1. Self-made dataset: It consists of 2100 images with helmet wearing targets, 1680 images as the training set and 420 images as the test set. It labels one category named "helmet" to test the single target detection effect of the algorithm. Labeling tool was used to label the dataset; the helmet area of each image was manually marked with a rectangular box and saved as YAML file.
2. GDU-HWD dataset: GDU-HWD has 3174 images in total, 2539 images as the training set and 635 images as the test set. Five detection categories are marked, namely, blue, yellow, white, red, and none. This dataset adds helmet color labels and negative samples, which can classify helmet colors to identify different types of work and improve the robustness of the model and reduce the error rate.

3.3 Experiment and Results

Evaluation Index

In the object detection task, it is needed not only to detect whether there is a target object in an image but also to find out the position of the object. It is necessary to take both precision and recall into consideration at the same time. Therefore, a standard for judging the quality of the training model – AP – is introduced. Conceptually, AP is the value of the region below the precision-recall curve:

$$\text{Precision} = \frac{TP}{TP+FP} \tag{3}$$

$$\text{Recall} = \frac{\mathrm{T}P}{TP+FN} \tag{4}$$

where TP is the number of helmets detected correctly, FP is the number of helmets misjudged, and FN is the number of missed detection.

It should be noticed that for multi-target detection, the concept of mAP needs to be introduced; mAP is the average value of AP of all categories, and "m" represents the number of categories.

Model Setting

YOLOv5 has four different depth networks: YOLOv5s, YOLOv5m, YOLOv5l, and YOLOv5x, corresponding to four pre-training weights. This experiment adopts a

Table 2 One target dataset

Method	Helmet	mAP
YOLO5	0.915	0.915
Enhanced YOLOv5	0.929	0.929

Table 3 Multi-target dataset

Method	Blue	White	Yellow	Red	None	mAP
Faster R-CNN [8]	0.708	0.680	0.696	0.609	0.590	0.656
SSD [8]	0.861	0.855	0.881	0.806	0.760	0.833
YOLOv5	0.947	0.945	0.868	0.850	0.817	0.885
EnhancedYOLOv5	0.940	0.954	0.895	0.892	0.829	0.902

lightweight YOLOv5s network model and pre-training weight named YOLOv5s.pt. Epoch was set to 100 rounds, batch size to 4, and initial learning rate to 0.001.

Result

Tables 2 and 3 are the model performance statistics obtained by training on two different datasets. The test results of the two datasets show that the detection accuracy of enhanced YOLOv5 is 1.4% and 1.7% higher than that of the original network, which proves that our improvements have a certain effect for detecting small targets such as safety helmets. On the GDUT-HWD dataset, through comparing with algorithms in other references, it can be found that YOLOv5 has a higher accuracy and the recognition speed of an image can reach 50FPS, which meets the requirements of real-time detection and has certain advantages in helmet detection work (Figs. 6 and 7).

Through the comparison of two datasets, it can be found that the accuracy of the single-target detection model is slightly higher than that of the multi-target detection model; this is because the multi-target dataset also classifies colors on the basis of the helmet, which increases the difficulty of detection. In practical engineering, increasing the color classification of helmets has more practical significance (Fig. 8).

Comparison of Helmet Test Results

In order to further verify the effectiveness of improvement, we selected the detection results of some complex scenes. As shown in the Figs. 9 and 10, the safety helmet appeared in the image as a small target. It can be seen that the improved YOLOV5 has better performance in complex scenes and effectively reduces the probability of missed detection, which indicates that the increased detection layer YOLOV5 can indeed improve the ability of detecting small targets and can be applied to various complex construction sites.

Fig. 6 Loss function curve

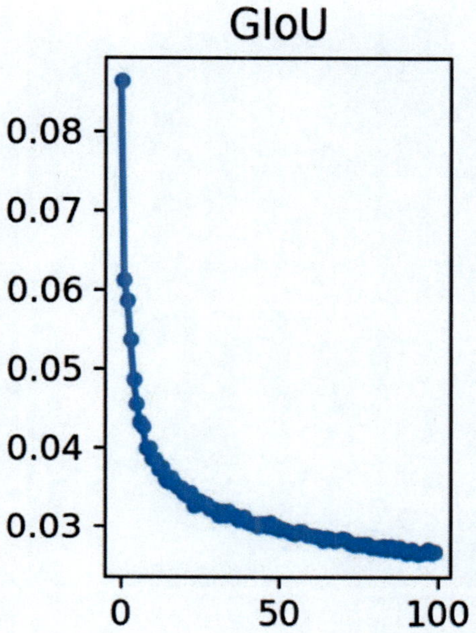

Fig. 7 mAP curve

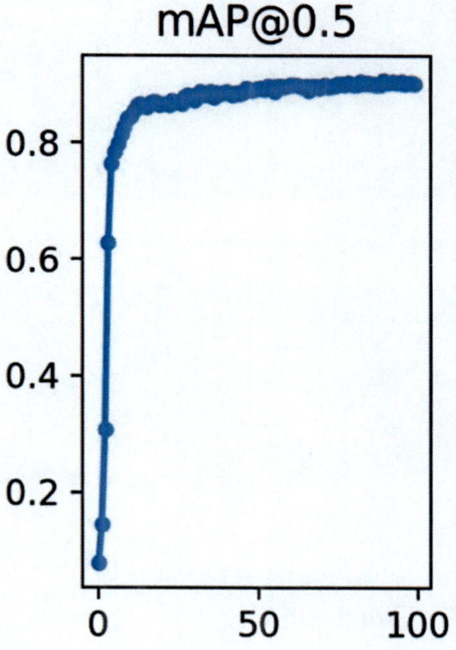

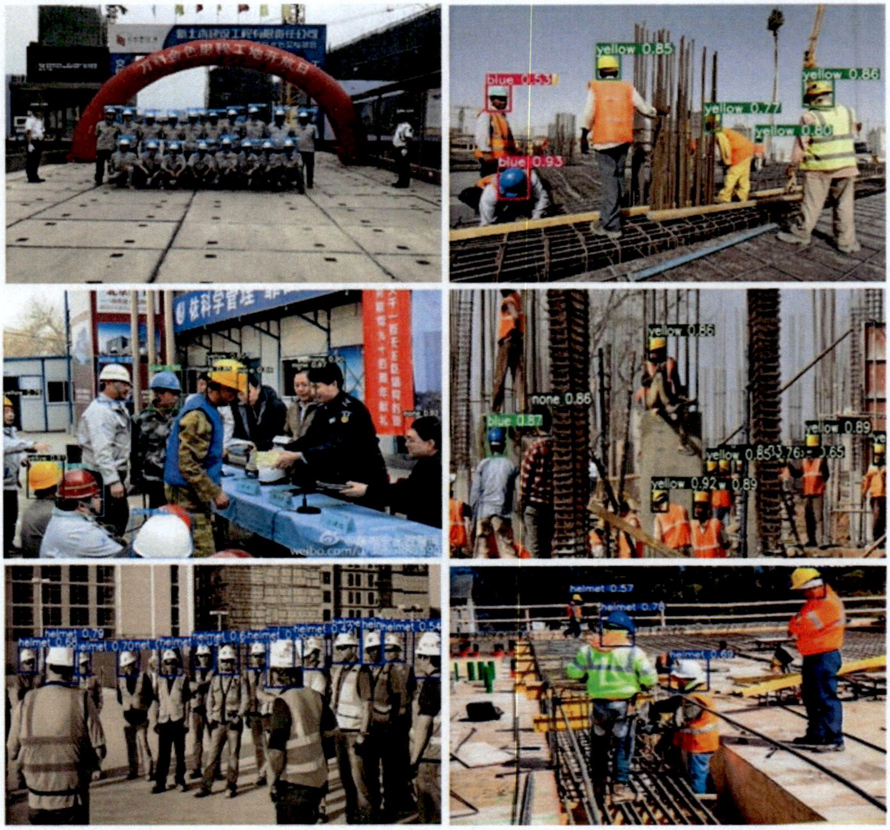

Fig. 8 Safety helmet detection result

Fig. 9 YOLOv5

4 Conclusion

Safety helmet wearing detection has great significances to ensure the safety of construction site workers. In this paper, YOLOv5 algorithm is used to realize the real-time detection of helmet wearing. On the basis of the original network, a prediction channel is added to improve the detection effect of small targets. The

Fig. 10 Enhanced YOLOv5

test results show that YOLOv5 can complete the detection work well. This paper has a certain effect in the modification of network structure. The image and video test results also achieve the effect of both accuracy and real-time performance. But in scenes with complex backgrounds and lots of shielding objects, there will still be misjudgments and missed detection. Consequently, it would be necessary to continue to optimize the algorithm and increase the number of datasets and the training samples. Based on the small size and real-time advantages of the YOLOv5 model, the optimized YOLOv5 algorithm can be deployed on the network in the future work and use cameras for real-time monitoring on the construction site. It can truly realize the safety helmet detection management system and accelerate the intelligent and humanized development of smart construction sites.

References

1. Li, J., Liu, H., Wang, T., Jiang, M.: Safety helmet wearing detection based on image processing and machine learning. In: 9th International Conference on Advanced Computational Intelligence (ICACI), pp. 201–205. Doha, Qator (2017)
2. Wu, X.: Innovative application of intelligent networking+big data in management of high-risk construction work. Build. Technol. Dev. **47**(2), 66–67 (2020)
3. Chen, C., Gong, W., Chen, Y., Li, W.: Learning a two-stage CNN model for multi-sized building detection in remote sensing images. Remote Sens. Lett. **10**(2), 103–110 (2019)
4. Altaf, K., Alexander, C., Hasan, D.: Image scene geometry recognition using low-level features fusion at multi-layer deep CNN. Neurocomputing. **440**, 111–126 (2021)
5. Feng, Z., Zhang, W., Zheng, Z.: Safety belt detection algorithm for aerial work based on mask R-CNN. Comput. Syst. Appl. **30**(3), 202–207 (2021)
6. Zhang, Y., Xu, X.: Safety helmet wearing detection method based on improved SSD. Electron. Meas. Technol. **43**(19), 80–84 (2020)
7. Mneymneh, B.E., Abbas, M., Khoury, H.: Vision-based framework for intelligent monitoring of hardhat wearing on construction sites. Comput. Civil Eng. **33** (2018). https://doi.org/10.1061/(ASCE)CP.1943-5487.0000813
8. Wu, J., Cai, N., Chen, W., Wang, H., Wang, G.: Automatic detection of hardhats worn by construction personnel: a deep learning approach and benchmark dataset. Autom. Constr. **106**, 102894 (2019)
9. Redmon, J., Farhadi, A.: YOLOv3: an incremental improvement. In: IEEE Conference on Computer Vision and Pattern Recognition (CVPR), pp. 18–22 (2018)

10. Wang, C.-Y., Liao, H.-Y.M., Wu, Y.-H., Chen, P.-Y., Hsieh, J.-W., Hau Yeh, I.: CSPNet: a new backbone that can enhance learning capability of CNN. In: Proceedings of the IEEE/CVF Conference on Computer Vision and Pattern Recognition Workshops, pp. 14–19 (2020)
11. Ren, S., He, K., Girshick, R., Sun, J.: Faster R-CNN: towards real-time object detection with region proposal networks. IEEE Trans. Pattern Anal. Mach. Intell. **39**(6), 1137–1149 (2017)
12. Girshick, R.: Fast R-CNN. In: IEEE International Conference on Computer Vision (2015)
13. Liu, W., Anguelov, D., Erhan, D., Szegedy, C., Reed, S., Cheng-Yang, F., Berg, A.C.: SSD single shot MultiBox detector. In: European Conference on Computer Vision. Springer, Cham (2016)
14. Redmon, J., Divvala, S., Girshick, R., Farhadi, A.: You only look once: unified, real-time object detection. In: Proceedings of the IEEE Conference on Computer Vision and Pattern Recognition (CVPR), Las Vegas, NV, USA (2016)

Index

D. Tang et al. (eds.), *Mobile Wireless Middleware, Operating Systems and Applications*, EAI/Springer Innovations in Communication and Computing,
https://doi.org/10.1007/978-3-030-98671-1